Geometry and Topology of Aspherical Manifolds

CONTEMPORARY MATHEMATICS

816

Geometry and Topology of Aspherical Manifolds

AMS Special Session
Singer–Hopf Conjecture in Geometry and Topology
March 18–19, 2023
Georgia Institute of Technology, Atlanta, GA

Luca F. Di Cerbo
Laurenţiu G. Maxim
Editors

AMS AMERICAN MATHEMATICAL SOCIETY

The frontispiece photos were used with permission under the Creative Commons Attribution-Share Alike 4.0 International license.

Hopf: By Ammann, Ernst – File: ETH–BIB–Hopf, Heinz (1894–1971)-Portr 07480.tif, CC BY-SA 4.0, `https://commons.wikimedia.org/w/index.php?curid=104005289`

Singer: By George M. Bergman, CC BY-SA 4.0, `https://commons.wikimedia.org/w/index.php?curid=110957295`

2020 *Mathematics Subject Classification.* Primary 14C30, 14F35, 14F45, 14J45, 20J05, 20J06, 57M07, 57N65, 58A10, 58A14.

Library of Congress Cataloging-in-Publication Data

Names: AMS Special Session on Singer–Hopf Conjecture in Geometry and Topology (2023 : Atlanta, Ga.) author. | Di Cerbo, Luca F. (Luca Fabrizio), 1982– editor. | Maxim, Laurenţiu G., editor.
Title: Geometry and topology of aspherical manifolds / AMS Special Sessionn on Singer–Hopf Conjecture in Geometry and Topology, March 18–19, 2023, Georgia Institute of Technology, Atlanta, Georgia ; Luca F. Di Cerbo, Laurenţiu G. Maxim, editors.
Description: Providence, Rhode Island : American Mathematical Society, [2025] | Series: Contemporary mathematics, 0271-4132 ; volume 816 | Includes bibliographical references.
Identifiers: LCCN 2024036013 | ISBN 9781470474959 (paperback) | ISBN 9781470478698 (ebook)
Subjects: LCSH: Topological manifolds–Congresses. | H-spaces–Congresses. | Sphere–Congresses. | AMS: Algebraic geometry – Cycles and subschemes – Transcendental methods, Hodge theory. | Algebraic geometry – (Co)homology theory – Homotopy theory; fundamental groups. | Algebraic geometry – (Co)homology theory – Topological properties. | Algebraic geometry – Surfaces and higher-dimensional varieties – Fano varieties. — Group theory and generalizations – Connections with homological algebra and category theory – Homological methods in group theory. | Group theory and generalizations – Connections with homological algebra and category theory – Cohomology of groups. | Manifolds and cell complexes – Low-dimensional topology – Topological methods in group theory. | Manifolds and cell complexes – Topological manifolds – Algebraic topology of manifolds. | Global analysis, analysis on manifolds – General theory of differentiable manifolds – Differential forms. | Global analysis, analysis on manifolds – General theory of differentiable manifolds – Hodge theory.
Classification: LCC QA613.2 .A49 2025 | DDC 514/.34–dc23/eng/20250207
LC record available at https://lccn.loc.gov/2024036013

DOI: https://doi.org/10.1090/conm/816

Color graphic policy. Any graphics created in color will be rendered in grayscale for the printed version unless color printing is authorized by the Publisher. In general, color graphics will appear in color in the online version.

∞ The paper used in this book is acid-free and falls within the guidelines
established to ensure permanence and durability.
Visit the AMS home page at `https://www.ams.org/`

10 9 8 7 6 5 4 3 2 1 30 29 28 27 26 25

Dedicated to the memory of Heinz Hopf (November 19, 1894–June 3, 1971)
and Isadore Manuel Singer (May 3, 1924–February 11, 2021)

Heinz Hopf

Isadore M. Singer

Contents

Preface

This volume has its origin in the American Mathematical Society (AMS) Special Session on the "Singer–Hopf Conjecture in Geometry and Topology", held during the Spring Southeastern Sectional Meeting from March 18 to March 19, 2023, hosted by the Georgia Institute of Technology in Atlanta, Georgia. This session featured 11 national and international speakers, and the talks were a mixture of presentations of research results and surveys and discussions of open problems. This structure is reflected in the contents of this volume that includes research and survey papers containing a substantial list of open problems and many interesting new research directions of current interest. Details about the conference, including a list of participants as well as short abstracts of the talks presented, are available from the 2023 AMS Spring Southeastern Sectional Meeting website:

`http://www.ams.org/meetings/sectional/2298_program_ss9.html#title`

This volume collects papers from speakers at the conference as well as other experts in the study of *aspherical manifolds* that were not able to attend this meeting. Indeed, the final scope of this volume is somewhat broader than the conference, hence the new title *Geometry and Topology of Aspherical Manifolds.*

Heinz Hopf was probably the first to investigate the connections between topology and curvature in a general and systematic way. Surprisingly, some of the questions that Hopf posed in the 1930s remain unanswered. A well-known example is the following problem on the sign of the Euler characteristic of aspherical manifolds. Recall that a manifold is said to be aspherical if its topological universal cover is contractible.

CONJECTURE 0.1 (Hopf). If X is a closed aspherical manifold of real dimension 2n, then

$$(-1)^n\chi_{top}(X) \geq 0.$$

Thanks to the uniformization theorem for Riemann surfaces, Hopf's conjecture is true when $n = 1$. On the other hand, this problem is still completely open even when $n = 2$! During the 1970s, Isadore Singer suggested an approach to this conjecture via the study of L^2-harmonic forms on the topological universal cover of X. In particular, Singer proposed the following conjecture whose resolution would settle the conjecture of Hopf.

CONJECTURE 0.2 (Singer). If X is a closed aspherical manifold of real dimension m, then

$$b_k^{(2)}(X;\widetilde{X}) = \begin{cases} (-1)^{m/2}\chi_{\mathrm{top}}(X) & \text{if} \quad k = m/2 \\ 0 & \text{if} \quad k \neq m/2, \end{cases}$$

where $\pi\colon \widetilde{X} \to X$ is the topological universal cover of X, and where the $b_k^{(2)}(X;\widetilde{X})$ are the Atiyah L^2-Betti numbers.

These problems motivated a substantial part of the work of Cheeger and Gromov in the 1980s, and they are now ubiquitous in many areas of modern mathematics. For example, they are widely studied in *geometric group theory,* where the study of normalized Betti numbers is currently a very active research area. They are also important in *geometric analysis* via the study of harmonic forms and the spectrum of the Hodge Laplacian. Finally, they play an important role in *complex algebraic geometry* as well, where they are intimately connected with the Shafarevich conjecture, the theory of the Albanese map, and various notions of hyperbolicity for algebraic varieties.

The thematic focus of our AMS Special Session was centered around these questions and our original goal was to provide an overview of several recent advances in the study of these long-standing problems. The goal of this volume is somewhat broader; indeed, it touches a larger set of problems concerning aspherical manifolds. For example, the survey by Kirsten, Kremer, and Lück covers a large spectrum of problems and conjectures concerning L^2-invariants of aspherical manifolds that are of interest not only in geometry and topology, but also in group theory and functional analysis. Remarkably, this survey covers exciting new developments on the $\mathbb{F}_p$-Singer conjecture by Avramidi, Okun, and Schreve, and it even includes a section exploring the connections between L^2-invariants and the theory condensed mathematics newly developed and championed by Clausen and Scholze. Similarly, the research article by Dranishnikov takes us outside the realm of the Hopf–Singer conjectures as it revisits the Lipschitz cohomology approach to the Novikov conjecture (developed by Connes, Gromov, and Moscovici in the early 1990s), and it includes new computations of the cohomology of large classes of aspherical manifolds. In a different direction, the survey paper by Kochloukova and Vidussi (with an appendix by Boggi) focuses on the finiteness properties of groups that can be written as extensions of finitely generated groups. Important examples of such groups come from the study of Kodaira fibrations and, more generally, from fibered Kähler groups. The authors conclude their survey with several applications of their results to the coherence of groups. The research paper by Arapura, Maxim, and Wang is dedicated to the study of the Hopf problem in algebraic geometry and Kähler geometry. It formulates several Hodge-theoretic extensions of the Hopf conjecture and presents a good amount of evidence, connecting different areas of mathematics such as topology, algebraic geometry, and differential geometry. We believe this contribution should be of interest to differential geometers and algebraic geometers alike, since it connects the circle of ideas of the Hopf conjecture together with the study of the positivity properties of the cotangent bundle of smooth projective varieties. This contribution is neatly complemented by the research and survey article by Liu, where some of these ideas are extended to smooth quasi-projective varieties. Liu's paper surveys recent developments about the asymptotic behavior of normalized Betti numbers for smooth quasi-projective varieties, and it is inspired by recent works of Arapura and Wang; Arapura, Maxim, and Wang; Di Cerbo and Lombardi; and Liu, Maxim, and Wang in the smooth projective setting. The volume concludes with two research papers of geometric analysis flavor. The paper by Di Cerbo and Hull focuses on the study of the Singer conjecture for generalized graph manifolds. In this setting, the Price equality for harmonic forms can be used in a flexible way thanks to the theory of collapsing of Cheeger, Gromov, and Fukaya. Among other things, the authors present a new proof of some classical vanishing results of Lott and Lück for aspherical 3-manifolds, and they also discuss the residual finiteness of the fundamental group of graph-like manifolds in dimensions $n \geq 4$. In the final contribution for this volume, Stern develops the theory of the Price equality for p-harmonic forms. The author uses these tools to recover some classical vanishing theorems of Pansu for the L_p-cohomology of Hadamard manifolds. The author concludes by addressing several cases of a conjecture of Gromov on the L_p-cohomology of symmetric spaces.

This volume presents a multidisciplinary point of view on the Singer conjecture, the Hopf conjecture, the study on normalized Betti numbers, and several other intriguing questions on the fundamental group and cohomology of aspherical manifolds. The problems are presented from several distinct points of view, and it is our hope that this volume will become a valuable resource for early career mathematicians interested in these deep and important questions. Indeed, this volume not only presents new and exciting results toward the solution of these questions, but also proposes extensions of such questions, clarifies their interdependence, and bridges a gap with closely related problems in differential geometry, complex algebraic geometry, and geometric topology. This volume highlights many interesting research directions in the study of aspherical manifolds and it covers a whole portfolio of problems and conjectures about L^2-invariants of aspherical manifolds. It provides a snapshot of contemporary research in mathematics at the interface of geometry and topology, and algebraic geometry. It is our hope that this volume will play an indirect role in focusing the attention of the mathematical community on these fascinating problems which continue to resist the siege of geometers and topologists.

The AMS publication department has been very supportive throughout the preparation of this book. We would like to thank Sergei Gelfand for noticing our AMS Special Session and for suggesting we embark in this project. Special thanks to Christine Thivierge for her guidance through each step of the preparation of this volume. The editors are grateful to Grigori Avramidi, Boris Goldfarb, Claudio Llosa Isenrich, Jean-Francois Lafont, Yongqiang Liu, Luigi Lombardi, Ismael Morales, Roman Sauer, Mark Stern, Robin Tucker-Drob, and Botong Wang for the help with several technical aspects of this volume.

This scholarship activity was partially supported by a grant of the *National Science Foundation* (DMS-2104662 to Luca F. Di Cerbo), by the *American Mathematical Society* (Meetings Child Care Grant to Luca F. Di Cerbo), and by the *Simons Foundation* (#567077 to Laurenţiu G. Maxim).

Luca Fabrizio Di Cerbo

Laurenţiu George Maxim

Survey and Research Articles

Contemporary Mathematics
Volume **816**, 2025
https://doi.org/10.1090/conm/816/16349

Some problems and conjectures about L^2-invariants

Dominik Kirstein, Christian Kremer, and Wolfgang Lück

ABSTRACT. In this paper we discuss open problems and conjectures concerning L^2-invariants.

1. Introduction

In this article we give a survey on open problems and conjectures concerning L^2-invariants. We cover the whole portfolio and not only certain aspects as they are considered in the previous more specialized (and within their scope more detailed) survey articles [**90, 92**]. Moreover, we include some new results and problems, which occurred after these two survey articles were written. The reader may select a specific topic by looking at the table of contents.

2. Basics about L^2-invariants

We present some basic definitions, constructions, and properties of L^2-invariants for the reader's convenience. The technical aspects of this section will not be needed in the other sections.

2.1. Group von Neumann algebras and their dimension function. The *group von Neumann algebra*

$$\mathcal{N}(G) := \mathcal{B}(L^2(G))^G \tag{2.1}$$

of the (discrete) group G is defined as the $\mathbb{C}$-algebra of G-equivariant bounded operators from $L^2(G)$ to $L^2(G)$, where $L^2(G)$ is the complex Hilbert space of square integrable functions from G to $\mathbb{C}$. It can also be viewed as a completion of the complex group ring $\mathbb{C}G$ with respect to the weak (or equivalently strong) operator topology. We will view it in the sequel just as a ring with unit and essentially forget all the functional analytic aspects. It has much nicer properties than $\mathbb{C}G$, for example, because it is *semi-hereditary*, i.e., any finitely generated submodule of a projective $\mathcal{N}(G)$-module is projective again. More information about von Neumann algebras and their basic properties can be found in [**60**, Section 2.2] and [**87**, Section 1.1 and Section 9.1].

2020 *Mathematics Subject Classification.* Primary 57Q99; Secondary 46L99, 20F99.
Key words and phrases. L^2-invariants, condensed sets.

The main feature is the *dimension function* that assigns to every $\mathcal{N}(G)$-module M an element

$$\dim_{\mathcal{N}(G)}(M) \in \mathbb{R}^{\geq 0} \amalg \{\infty\}, \tag{2.2}$$

for $\mathbb{R}^{\geq 0} = \{r \in \mathbb{R} \mid r \geq 0\}$. We summarize its main properties:

(1) It is *additive* on short exact sequences of $\mathcal{N}(G)$-modules, also in the extreme cases following the convention that for $x \in \mathbb{R}^{\geq 0} \amalg \{\infty\}$ we set $x + \infty = \infty$.
(2) The dimension function extends the classical Murray–von Neumann dimension for finitely generated projective $\mathcal{N}(G)$-modules defined by the Hattori–Stalling rank associated to the standard trace of $\mathcal{N}(G)$. In particular, we have $\dim_{\mathcal{N}(G)}(\mathcal{N}(G)) = 1$, which, together with additivity, implies that the dimension of a finitely generated $\mathcal{N}(G)$-module is finite. Note that the converse is not true; see [**87**, Example 6.12 on page 242].
(3) It is *cofinal* in the sense that for any directed union $M = \bigcup_{i\in I} M_i$ of $\mathcal{N}(G)$-modules the equality $\dim_{\mathcal{N}(G)}(M) = \sup\{\dim_{\mathcal{N}(G)}(M_i) \mid i \in I\}$ holds.
(4) It satisfies the *continuity property*. A special case says that for any finitely generated $\mathcal{N}(G)$-module M one can find a finitely generated projective $\mathcal{N}(G)$-module PM together with two exact sequences $0 \to TM \to M \to PM \to 0$ and $0 \to \mathcal{N}(G)^n \to \mathcal{N}(G)^n \to TM \to 0$ for which $\dim_{\mathcal{N}(G)}(TM) = 0$ and $\dim_{\mathcal{N}(G)}(M) = \dim_{\mathcal{N}(G)}(PM)$ hold. The general version can be found in [**87**, Theorem 6.7].
(5) We have $\dim_{\mathcal{N}(G)}(M) = 0$ if and only if every projective submodule of M is trivial. Moreover, a finitely generated projective $\mathcal{N}(G)$-module M is trivial if and only if $\dim_{\mathcal{N}(G)}(M) = 0$ holds.

It turns out that the dimension function is uniquely determined by the properties (1)–(4). In the special case where G is finite and M is an $\mathcal{N}(G)$-module one has $\mathcal{N}(G) = \mathbb{C}G$, and $|G| \cdot \dim_{\mathcal{N}(G)}(M)$ is the (classical) dimension of the underlying complex vector space of M. If G contains an element of infinite order, then every element $r \in \mathbb{R}^{\geq 0}$ can be realized as $r = \dim_{\mathcal{N}(G)}(M)$ for some finitely generated projective $\mathcal{N}(G)$-module M. More information about the dimension function can be found in [**60**, Sections 2.3 and 4.2], [**85**], [**86**], [**87**, Chapter 6].

2.2. L^2-Betti numbers. Let Y be a G-space. Its singular chain complex $C_*^s(Y)$ is acted upon by G, so it acquires the structure of a chain complex of $\mathbb{Z}G$-modules. Note that $\mathcal{N}(G)$ can be viewed as an $\mathcal{N}(G)$-$\mathbb{Z}G$-bimodule. Then we obtain an $\mathcal{N}(G)$-chain complex $\mathcal{N}(G) \otimes_{\mathbb{Z}G} C_*^s(X)$. Its nth homology $H_n(\mathcal{N}(G) \otimes_{\mathbb{Z}G} C_*^s(X))$ is a $\mathcal{N}(G)$-module. We define the *nth-L^2-Betti number* of the G-space Y to be

$$b_n^{(2)}(Y; \mathcal{N}(G)) := \dim_{\mathcal{N}(G)}\big(H_n(\mathcal{N}(G) \otimes_{\mathbb{Z}G} C_*^s(Y))\big) \in \mathbb{R}^{\geq 0} \amalg \{\infty\}. \tag{2.3}$$

If Y is a G-CW-complex, we can replace $C_*^s(Y)$ by the cellular $\mathbb{Z}G$-chain complex $C_*(Y)$ in (2.3). If Y is a G-CW-complex such that Y/G is a finite d-dimensional CW-complex, then $b_n^{(2)}(Y; \mathcal{N}(G))$ is finite for every $n \geq 0$ and vanishes for $n > d$.

We are mainly be interested in the case where X is a path connected space possessing a universal covering $p\colon \widetilde{X} \to X$. Recall that $\widetilde{X}$ then comes with a free π-action for π the fundamental group of X, and we abbreviate

$$b_n^{(2)}(\widetilde{X}) = b_n^{(2)}(\widetilde{X}; \mathcal{N}(\pi)) \in \mathbb{R}^{\geq 0} \amalg \{\infty\}. \tag{2.4}$$

If X is not path connected, one defines $b_n^{(2)}(\widetilde{X})$ to be $\sum_{C\in\pi_0(X)} b_n^{(2)}(\widetilde{C})$. If X is a CW-complex of finite type, then $b_n^{(2)}(\widetilde{X})$ is finite. If M is a closed Riemannian manifold, then $b_n^{(2)}(\widetilde{M})$ agrees with the original definition of the L^2-Betti number due to Atiyah [**9**] in terms of heat kernel on the universal covering, namely, we have

$$b_n^{(2)}(\widetilde{M}) = \lim_{t\to\infty} \int_{\mathcal{F}} \operatorname{tr}(e^{-t\widetilde{\Delta}_n}(\widetilde{x},\widetilde{x}))\,\mathrm{dvol} \tag{2.5}$$

where $\mathcal{F} \subseteq \widetilde{X}$ is a fundamental domain for the π-action on $\widetilde{M}$ and $\operatorname{tr}(e^{-t\widetilde{\Delta}_n}(\widetilde{x},\widetilde{x}))$ is the trace of the heat kernel on $\widetilde{M}$ at $(\widetilde{x},\widetilde{x})$ for $\widetilde{x} \in \widetilde{X}$ at the time t in degree n.

If G is a (discrete) group, we define its *nth-L^2-Betti number*

$$b_n^{(2)}(G) := b_n^{(2)}(EG;\mathcal{N}(G)) \in \mathbb{R}^{\geq 0} \amalg \{\infty\}. \tag{2.6}$$

where EG is a the total space of the universal principal G-bundle $EG \to BG$. If $\underline{E}G$ is the classifying space for proper G-actions—see, for instance, [**88**]—then we obtain $b_n^{(2)}(G) = b_n^{(2)}(\underline{E}G;\mathcal{N}(G))$ from [**87**, Theorem 6.54 (2) on page 265]. This is useful since there are often small models of $\underline{E}G$ available when EG may have no such nice models. For example, when G has torsion every CW-model for EG has cells in infinitely many dimension, whereas one might construct a finite dimensional model for $\underline{E}G$ to conclude vanishing of L^2-Betti numbers in high degrees.

Roughly speaking, the usefulness of the L^2-Betti numbers stems often from the fact that they vanish quite often and pose interesting obstructions to the solution of geometric problems if they do not. For instance, for a connected CW-complex X we have $b_0(X) = 1$, whereas $b_0^{(2)}(\widetilde{X})$ vanishes if π is infinite and is equal to $|\pi|^{-1}$ if π is finite. We mention as an appetizer two rather surprising examples.

EXAMPLE 2.1. If the group G contains an amenable infinite normal subgroup $H \subseteq G$, then $b_n^{(2)}(G)$ vanishes for every $n \geq 0$. This implies that the Euler characteristic $\chi(BG)$ vanishes, if we additionally assume that BG has a finite CW-model. Note that we do not require anything about BH. This is a special case of [**87**, Theorem 7.2].

EXAMPLE 2.2. If $1 \to K \to G \to Q \to 1$ is an extension of infinite groups such that K is finitely generated and G is finitely presented, then $b_1^{(2)}(G)$ vanishes and hence the deficiency of G is bounded from above by 1. Recall that the deficiency is the maximum of the set of integers $g(P) - r(P)$, where P runs through all finite presentations P of G and $g(P)$ is the number of generators and $r(P)$ the number of relations. [**83**, Theorem 6.1]

2.3. Fuglede–Kadison determinants. Given an $\mathcal{N}(G)$-map $f\colon P \to Q$ of finitely generated projective $\mathcal{N}(G)$-modules, one can define its *Fuglede–Kadison determinant*

$$\det_{\mathcal{N}(G)}(f) \in \mathbb{R}^{\geq 0} := \{r \in \mathbb{R} \mid r \geq 0\}. \tag{2.7}$$

Let us briefly explain its definition, even though it will be of no concern to us later. One can interprete f as a bounded G-equivariant operator $f\colon V \to W$ of *finitely generated Hilbert $\mathcal{N}(G)$-modules*, i.e., V is a Hilbert space V together with a linear isometric G-action such that there exists an isometric linear G-embedding of V into $L^2(G)^r$ for some natural number r, and analogously for W. Denote by

$\{E_\lambda^{f^*f} \mid \lambda \in \mathbb{R}\}$ the (right-continuous) family of spectral projections of the positive operator $f^*f \colon V \to V$. Define the *spectral density function of* f by

$$F_f \colon \mathbb{R} \to \mathbb{R}^{\geq 0} \quad \lambda \mapsto \dim_{\mathcal{N}(G)}\big(\operatorname{im}(E_{\lambda^2}^{f^*f})\big).$$

The spectral density function is monotone, nondecreasing, right-continuous and satisfies $F(0) = \dim_{\mathcal{N}(G)}(\ker(f))$. Let dF be the unique measure on the Borel σ-algebra on $\mathbb{R}$ that satisfies $dF((a,b]) = F(b) - F(a)$ for $a < b$. Define Fuglede–Kadison determinant $\det_{\mathcal{N}(G)}(f) \in \mathbb{R}^{\geq 0}$ to be

$$\det_{\mathcal{N}(G)}(f) = \exp\left(\int_{0+}^{\infty} \ln(\lambda)\, dF\right)$$

if the Lebesgue integral $\int_{0+}^{\infty} \ln(\lambda)\, dF$ is a real number, and to be 0 if we have $\int_{0+}^{\infty} \ln(\lambda)\, dF = -\infty$. We say that f is of *determinant class* if $\det(f) \neq 0$. With our conventions we have $\det(0) = 1$ for the zero map $0 \colon P \to Q$.

EXAMPLE 2.3 (Fuglede–Kadison determinant for finite groups). To illustrate this definition, we look at the example where G is finite. Then $\mathcal{N}(G)$ is the same as $\mathbb{C}G$ and we can think of f as a $\mathbb{C}G$-linear map of finite-dimensional unitary G-representations. The spectral density function F_f is the right-continuous step function, whose value at λ is the sum of the complex dimensions of the eigenspaces of f^*f for eigenvalues $\mu \leq \lambda^2$ divided by the order of G. Let $\lambda_1, \lambda_2, \ldots, \lambda_r$ be the nonzero eigenvalues of f^*f with multiplicity μ_i. One computes

$$\det_{\mathcal{N}(G)}(f) \quad = \quad \exp\left(\sum_{i=1}^{r} \frac{\mu_i}{|G|} \cdot \ln(\sqrt{\lambda_i})\right) \quad = \quad \prod_{i=1}^{r} \lambda_i^{\frac{\mu_i}{2\cdot|G|}} \quad = \quad \det_{\mathbb{C}}\big(\overline{f^*f}\big)^{\frac{1}{2\cdot|G|}}.$$

Here, $\overline{f^*f}$ denotes the automorphism of the orthogonal complement of the kernel of f^*f induced by f^*f and $\det_{\mathbb{C}}\big(\overline{f^*f}\big)$ is put to be 1 if f is the zero operator. If $f \colon \mathbb{C}G^m \to \mathbb{C}G^m$ is an automorphism, we get

$$\det_{\mathcal{N}(G)}(f) = |\det_{\mathbb{C}}(f)|^{\frac{1}{|G|}}.$$

More information about the Fuglede–Kadison determinant can be found for instance in [**87**, Section 3.2].

2.4. L^2-torsion. Let X be a connected finite CW-complex. Let $C_*(\widetilde{X})$ be the cellular $\mathbb{Z}\pi$-chain complex of its universal covering for π the fundamental group of X. Then $\mathcal{N}(\pi) \otimes_{\mathbb{Z}\pi} C_*(\widetilde{X})$ is an $\mathcal{N}(\pi)$-chain complex, which is finite dimensional and consists of finitely generated free $\mathcal{N}(\pi)$-modules. If $c_n \colon C_n(\widetilde{X}) \to C_{n-1}(\widetilde{X})$ is the nth differential of $C_*(\widetilde{X})$, then

$$\operatorname{id}_{\mathcal{N}(\pi)} \otimes_{\mathbb{Z}\pi} c_n \colon \mathcal{N}(\pi) \otimes_{\mathbb{Z}\pi} C_n(\widetilde{X}) \to \mathcal{N}(\pi) \otimes_{\mathbb{Z}\pi} C_{n-1}(\widetilde{X})$$

is an $\mathcal{N}(\pi)$-map of finitely generated free $\mathcal{N}(\pi)$-modules and its Fuglede–Kadison determinant $\det_{\mathcal{N}(\pi)}(\operatorname{id}_{\mathcal{N}(\pi)} \otimes_{\mathbb{Z}\pi} c_n) \in \mathbb{R}^{\geq 0}$ is defined. We say that X is *of determinant class* if $\det_{\mathcal{N}(\pi)}\big(\operatorname{id}_{\mathcal{N}(\pi)} \otimes_{\mathbb{Z}\pi} c_n\big) \neq 0$ holds for every $n \geq 1$, and in this case we define the *L^2-torsion* of X to be

$$\rho^{(2)}(\widetilde{X}) = -\sum_{n \geq 1} (-1)^n \cdot \ln\big(\det_{\mathcal{N}(\pi)}(\operatorname{id}_{\mathcal{N}(\pi)} \otimes_{\mathbb{Z}\pi} c_n)\big) \in \mathbb{R}. \tag{2.8}$$

We say that X is *L^2-acyclic* if $b_n^{(2)}(\widetilde{X}) = 0$ for $n \geq 0$. We say that X is *det-L^2-acyclic* if it is L^2-acyclic and of determinant class. If X is not path connected, we define $\rho^{(2)}(\widetilde{X}) = \sum_{C \in \pi_0(X)} \rho^{(2)}(\widetilde{C})$.

The definition of L^2-torsion in the analytic setting goes back to Lott [**80**] and Mathai [**100**], and in the topological setting to Lück–Rothenberg [**95**]. The equality of these two versions has been proved in [**21**].

2.5. Basic properties of L^2-Betti numbers and L^2-torsion. Here is a list of basic properties of L^2-Betti numbers and L^2-torsion.

THEOREM 2.4.

(1) (Simple) Homotopy invariance, *see* [**87**, Theorem 1.35 (1) on page 37 and Theorem 3.96 (i) on page 163],[**91**, Theorem 6.7 (2)].
   (a) *If X and Y are homotopy equivalent spaces, then we get for $n \geq 0$*
   $$b_n^{(2)}(\widetilde{X}) = b_n^{(2)}(\widetilde{Y});$$
   (b) *Suppose that X and Y are homotopy equivalent finite CW-complexes and $\widetilde{X}$ or $\widetilde{Y}$ is* det-L^2*-acyclic. Then both $\widetilde{X}$ and $\widetilde{Y}$ are* det-L^2*-acyclic;*
   (c) *Let $f\colon X \to Y$ be a homotopy equivalence of finite CW-complexes. Assume that $\widetilde{X}$ and $\widetilde{Y}$ are* det-L^2*-acyclic. Suppose that f is simple, or that $\pi_1(X)$ satisfies the Determinant Conjecture* 5.1*, or that the K-theoretic Farrell–Jones Conjecture for $\mathbb{Z}\pi_1(X)$ holds. Then*
   $$\rho^{(2)}(\widetilde{Y}) = \rho^{(2)}(\widetilde{X});$$

(2) Euler–Poincaré formula*; see* [**87**, Theorem 1.35 (2) on page 37].
We get for the Euler characteristic $\chi(X)$ of a finite CW-complex X
$$\chi(X) = \sum_{n \geq 0} (-1)^n \cdot b_n^{(2)}(\widetilde{X});$$

(3) Sum formula*; see* [**87**, Theorem 3.96 (2) on page 164].
Consider a pushout of finite CW-complexes such that j_1 is an inclusion of CW-complexes, j_2 is cellular, and X inherits its CW-complex structure from X_0, X_1 and X_2
$$\begin{array}{ccc} X_0 & \xrightarrow{j_1} & X_1 \\ {\scriptstyle j_2}\downarrow & & \downarrow{\scriptstyle i_1} \\ X_2 & \xrightarrow[i_2]{} & X. \end{array}$$
Assume that for $k = 0, 1, 2$ the map $\pi_1(i_k, x)\colon \pi_1(X_k, x) \to \pi_1(X, j_k(x))$ induced by the obvious map $i_k\colon X_k \to X$ is injective for all base points x in X_k.
   (a) *If $\widetilde{X_0}$, $\widetilde{X_1}$, and $\widetilde{X_2}$ are L^2-acyclic, then $\widetilde{X}$ is L^2-acyclic;*
   (b) *If $\widetilde{X_0}$, $\widetilde{X_1}$, and $\widetilde{X_2}$ are* det-L^2*-acyclic, then $\widetilde{X}$ is* det-L^2*-acyclic and we get*
   $$\rho^{(2)}(\widetilde{X}) = \rho^{(2)}(\widetilde{X_1}) + \rho^{(2)}(\widetilde{X_2}) - \rho^{(2)}(\widetilde{X_0});$$

(4) Poincaré duality*; see* [**87**, Theorem 1.35(3) on p. 37 and Theorem 3.96(3) on p. 164].
Let M be a closed manifold of dimension d

(a) *Then*

$$b_n^{(2)}(\widetilde{M}) = b_{d-n}^{(2)}(\widetilde{M});$$

(b) *Suppose that n is even and $\widetilde{M}$ is* det-L^2*-acyclic. Then*

$$\rho^{(2)}(\widetilde{M}) = 0;$$

(5) Product formula*; see* [**87**, Theorem 1.35(4) on p. 37 and Theorem 3.96(4) on p. 164].
Let X and Y be finite CW-complexes.
(a) *Then*

$$b_n^{(2)}(\widetilde{X \times Y}) = \sum_{\substack{i,j \in \mathbb{N} \\ n=i+j}} b_i^{(2)}(\widetilde{X}) \cdot b_j^{(2)}(\widetilde{Y});$$

(b) *Suppose that $\widetilde{X}$ is* det-L^2*-acyclic. Then $\widetilde{X \times Y}$ is* det-L^2*-acyclic and*

$$\rho^{(2)}(\widetilde{X \times Y}) = \chi(Y) \cdot \rho^{(2)}(\widetilde{X});$$

(6) Multiplicativity*, see* [**87**, Theorem 1.35(9) on p. 38 and Theorem 3.96(5) on p. 164].
Let $X \to Y$ be a finite covering of finite CW-complexes with d sheets.
(a) *Then we get for $n \geq 0$*

$$b_n^{(2)}(\widetilde{X}) = d \cdot b_n^{(2)}(\widetilde{Y});$$

(b) *Then $\widetilde{X}$ is* det-L^2*-acyclic if and only if $\widetilde{Y}$ is* det-L^2*-acyclic, and in this case*

$$\rho^{(2)}(\widetilde{X}) = d \cdot \rho^{(2)}(\widetilde{Y});$$

(7) Determinant class.
If $\pi_1(C)$ satisfies the Determinant Conjecture 5.1 *for each component C of the finite CW-complex X, then $\widetilde{X}$ is of determinant class;*
(8) 0th L^2-Betti number*, see* [**87**, Theorem 1.35 (8) on p. 38].
If X is a connected finite CW-complex with fundamental group π, then

$$b_0^{(2)}(\widetilde{X}) = \begin{cases} \frac{1}{|\pi|} & \text{if } \pi \text{ is finite;} \\ 0 & \text{otherwise;} \end{cases}$$

(9) Fibration formula*, see* [**87**, Lemma 1.41 on p. 45 and Corollary 3.103 on p. 166].
(a) *Let $p\colon E \to B$ a fibration whose base space B is a connected finite CW-complex and whose fiber is homotopy equivalent to a finite CW-complex Z. Suppose that for every $b \in B$ and $x \in F_b := p^{-1}(b)$ the inclusion $p^{-1}(b) \to E$ induces an injection on the fundamental groups $\pi_1(F_b, x) \to \pi_1(E, x)$, and that Z is L^2-acyclic.*
Then E is homotopy equivalent to a finite CW-complex X which is L^2-acyclic;
(b) *Let $F \xrightarrow{i} E \xrightarrow{p} B$ be locally trivial fiber bundle of finite CW-complexes. Suppose that B is connected, the map $\pi_1(F, x) \to \pi_1(E, i(x))$ is bijective for every base point $x \in F$, and $\widetilde{F}$ is* det-L^2*-acyclic.*
Then $\widetilde{E}$ is det-L^2*-acyclic and*

$$\rho^{(2)}(\widetilde{E}) = \chi(B) \cdot \rho^{(2)}(\widetilde{F});$$

(10) S^1-actions; *see* [**87**, Theorem 1.40 on p. 43 and Theorem 3.105 on p. 168]. *Let X be a connected compact S^1-CW-complex, for instance, a closed smooth manifold with smooth S^1-action. Suppose that for one orbit S^1/H (and hence for all orbits) the inclusion into X induces a map on π_1 with infinite image (so, in particular, the S^1-action has no fixed points). Then $\widetilde{X}$ is* det-L^2*-acyclic and $\rho^{(2)}(\widetilde{M})$ vanishes;*

(11) Aspherical spaces; *see* [**87**, Theorem 3.111 on p. 171 and Theorem 3.113 on p. 172].

   (a) *Let M be an aspherical closed smooth manifold with a smooth S^1-action. Then the conditions appearing in assertion* (10) *are satisfied and hence $\widetilde{M}$ is* det-L^2*-acyclic and $\rho^{(2)}(\widetilde{X})$ vanishes;*
   (b) *If X is an aspherical finite CW-complex whose fundamental group contains an elementary amenable infinite normal subgroup, then $\widetilde{X}$ is* det-L^2*-acyclic and $\rho^{(2)}(\widetilde{X})$ vanishes;*

(12) Mapping tori; *see* [**87**, Theorem 1.39 on p. 42]. *Let $f\colon X \to X$ be a self homotopy equivalence of a finite CW-complex. Denote by T_f its mapping torus.*

   (a) *Then $\widetilde{T_f}$ is L^2-acyclic;*
   (b) *If $\widetilde{X}$ is* det-L^2*-acyclic, then $\rho^{(2)}(\widetilde{T_f})$ vanishes;*

(13) Hyperbolic manifolds; *see* [**52**], [**87**, Theorem 1.39 on p. 42]. *Let M be a hyperbolic closed manifold of dimension d.*

   (a) *If d is odd, $\widetilde{M}$ is* det-L^2*-acyclic;*
   (b) *Suppose that $d = 2m$ is even. Then $b_n^{(2)}(\widetilde{M})$ vanishes for $n \neq m$, and we have $(-1)^m \cdot \chi(M) = b_m^{(2)}(\widetilde{M}) > 0$;*
   (c) *For every number m there exists an explicit constant $C_m > 0$ with the following property: If M is a hyperbolic closed manifold of dimension $(2m+1)$ with volume $\mathrm{vol}(M)$, then*

$$\rho^{(2)}(\widetilde{M}) = (-1)^m \cdot C_m \cdot \mathrm{vol}(M).$$

   We have $C_1 = \frac{1}{6\pi}$. The number $\pi^m \cdot C_m$ is always rational;

(14) Approximation of L^2-Betti numbers by classical Betti numbers; *see* [**82**], [**87**, Chapter 13]. *Let X be a connected finite CW-complex with fundamental group $G = \pi_1(X)$. Suppose that G comes with a descending chain of subgroups*

$$G = G_0 \supseteq G_1 \supseteq G_2 \supseteq \cdots$$

such that G_i is normal in G, the index $[G : G_i]$ is finite, and we have $\bigcap_{i \geq 0} G_i = \{1\}$.

Then $G_i \backslash \widetilde{X} \to X$ is a finite $[G : G_i]$-sheeted covering, and we get for $n \geq 0$

$$b_n^{(2)}(\widetilde{X}) = \lim_{i \to \infty} \frac{b_n(G_i \backslash \widetilde{X})}{[G : G_i]},$$

where $b_n(G_i \backslash \widetilde{X})$ is the classical nth Betti number of the finite CW-complex $G_i \backslash \widetilde{X}$.

For more information about L^2-invariants and their applications to algebra, geometry, group theory, index theory, operator algebras, topology, and K-theory, for instance, we refer the reader to **[60]** and **[87]**. We will discuss some of them later in this article.

3. The Atiyah Conjecture

In this section we discuss one of the most prominent conjectures in the field of L^2-invariants, the Atiyah Conjecture. It predicts the possible values of L^2-betti numbers and has many interesting implications. For example, Kaplansky's Zero Divisor Conjecture for rational group rings follows from it.

3.1. Statement of the Atiyah Conjecture.

CONJECTURE 3.1 (Atiyah Conjecture). *Consider a field F with $\mathbb{Q} \subseteq F \subseteq \mathbb{C}$. Let G be a group possessing an upper bound on the orders of its finite subgroups. Let* $\operatorname{lcm}(G)$ *be the natural number given by the least common multiple of the orders of its finite subgroups.*

We say that G satisfies the Atiyah Conjecture with coefficients in F, *if for any finitely presented FG-module M the von Neumann dimension satisfies*

$$\operatorname{lcm}(G) \cdot \dim_{\mathcal{N}(G)}(\mathcal{N}(G) \otimes_{FG} M) \in \mathbb{Z}.$$

Note that G is torsion-free if and only if $\operatorname{lcm}(G) = 1$, so the Atiyah conjecture in this case predicts integrality of L^2-Betti numbers. For a field F, let us denote by $\mathcal{A}_F$ the collection of all groups which admit a bound on the order of their finite subgroups and satisfy the Atiyah Conjecture with coefficients in F. Obviously, $\mathcal{A}_{\mathbb{C}} \subseteq \mathcal{A}_F \subseteq \mathcal{A}_{\mathbb{Q}}$.

REMARK 3.2 (Equivalent formulations of the Atiyah Conjecture 3.1). Consider a field F with $\mathbb{Q} \subseteq F \subseteq \mathbb{C}$. Let G be a group that possesses an upper bound on the orders of is finite subgroups. Then the following assertions are equivalent:

(1) We have $G \in \mathcal{A}_F$;
(2) For any matrix $A \in M_{m,n}(FG)$ the dimension $\dim_{\mathcal{N}(G)}(\ker(r_A))$ of the kernel of the $\mathcal{N}(G)$-homomorphism $r_A \colon \mathcal{N}(G)^m \to \mathcal{N}(G)^n$ given by right multiplication with A satisfies $\operatorname{lcm}(G) \cdot \dim_{\mathcal{N}(G)}(\ker(r_A)) \in \mathbb{Z}$;
(3) For any FG-chain complex C_* of finitely generated free FG-modules and all $n \in \mathbb{Z}$ we have
$$\operatorname{lcm}(G) \cdot \dim_{\mathcal{N}(G)}(H_n(\mathcal{N}(G) \otimes_{FG} C_*)) \in \mathbb{Z}.$$

If we additionally assume $F = \mathbb{Q}$, then the three assertions above and the following assertion are equivalent:

(4) For any $\mathbb{Z}G$-chain complex C_* of finitely generated free $\mathbb{Z}G$-modules and all $n \in \mathbb{Z}$ we have
$$\operatorname{lcm}(G) \cdot \dim_{\mathcal{N}(G)}(H_n(\mathcal{N}(G) \otimes_{\mathbb{Z}G} C_*)) \in \mathbb{Z}.$$

If we additionally assume $F = \mathbb{Q}$ and that G is finitely presented, then the four assertions above and the following assertion are equivalent:

(5) For any closed manifold M with $\pi_1(M) \cong G$ and any $n \geq 0$ we have
$$\operatorname{lcm}(G) \cdot b_n^{(2)}(\widetilde{M}) \in \mathbb{Z}.$$

All of these claims follow from from [**87**, Lemma 10.5 on p. 371 and Lemma 10.7 on p. 372], the equality $\dim_{\mathcal{N}(G)}(\mathcal{N}(G)) = 1$, and the additivity of the dimension function $\dim_{\mathcal{N}(G)}$.

The Atiyah Conjecture 3.1 is rather surprizing in view of (2.5) and assertion (5) appearing in Remark 3.2, since there seems to be no reason why the expression appearing on the right-hand side of (2.5) should be an integer if $\pi_1(M)$ is torsionfree.

3.2. Status of the Atiyah Conjecture. The notions of elementary amenable groups and amenable groups are explained in [**87**, Subsection 6.4.1].

DEFINITION 3.3 (Class of groups $\mathcal{C}$). Let $\mathcal{C}$ be the smallest class of groups satisfying the following conditions:

(1) $\mathcal{C}$ contains all free groups;
(2) If $\{G_i \mid i \in I\}$ is a directed system of subgroups directed by inclusion such that each G_i belongs to $\mathcal{C}$, then $G = \bigcup_{i\in I} G_i$ belongs to $\mathcal{C}$;
(3) Let $1 \to K \to G \to Q \to 1$ be an extension of groups such that K belongs to $\mathcal{C}$ and Q is elementary amenable, then G belongs to $\mathcal{C}$.

DEFINITION 3.4 (Class of groups $\mathcal{D}$). Let $\mathcal{D}$ be the smallest class of groups satisfying the following conditions:

(1) The trivial group belongs to $\mathcal{D}$;
(2) If $\{G_i : i \in I\}$ is a filtered system of groups in $\mathcal{D}$ (with arbitrary structure maps), then its colimit again belongs to $\mathcal{D}$;
(3) If $\{G_i : i \in I\}$ is a cofiltered system of groups in $\mathcal{D}$ (with arbitrary structure maps), then its limit again belongs to $\mathcal{D}$;
(4) If G belongs to $\mathcal{D}$ and $H \subseteq G$ is a subgroup, then $H \in \mathcal{D}$;
(5) If $p\colon G \to A$ is an epimorphism of a torsionfree group G onto an elementary amenable group A and if $p^{-1}(B) \in \mathcal{A}_F$ for every finite group $B \subset A$, then $G \in \mathcal{A}_F$.

Note that each element in $\mathcal{D}$ is a torsion-free group, and the class $\mathcal{D}$ contains all residually (torsion-free elementary amenable) groups.

A group is called *locally indicable*, if every nontrivial finitely generated subgroup admits an epimorphism onto $\mathbb{Z}$. Locally indicable groups are torsionfree. Examples for locally indicable groups are torsion-free one-relator groups.

THEOREM 3.5 (Status of the Atiyah Conjecture 3.1 with coefficients in F). *Consider a field F with $\mathbb{Q} \subseteq F \subseteq \mathbb{C}$.*

(1) *If G belongs to $\mathcal{C}$ and possesses an upper bound on the orders of is finite subgroups, then $G \in \mathcal{A}_F$;*
(2) *If G belongs to $\mathcal{D}$, then G is torsion-free and $G \in \mathcal{A}_F$;*
(3) *If $1 \to H \to G \to Q \to 1$ is an extension of groups, H is torsion-free and belongs to $\mathcal{A}_F$, and Q is locally indicable, then G is torsion-free and belongs to $\mathcal{A}_F$;*
(4) *If $G \in \mathcal{A}_F$ and $H \subseteq G$ is a subgroup, then $H \in \mathcal{A}_F$;*
(5) *If G is the directed union $\bigcup_{i\in I} G_i$ of subgroups G_i directed by inclusion and each G_i belongs to $\mathcal{A}_F$, then G belongs to $\mathcal{A}_F$;*
(6) *The group G belongs to $\mathcal{A}_F$ if and only if all its finitely generated subgroups belong to $\mathcal{A}_F$;*

(7) *If $1 \to K \to G \to Q \to 1$ is an extension of groups such that K is finite and G belongs to $\mathcal{A}_F$, then Q belongs to $\mathcal{A}_F$;*
(8) *Let M be a connected (not necessarily compact) d-dimensional manifold (possibly with nonempty boundary) such that $d \leq 3$ and its fundamental group $\pi_1(M)$ is torsion-free, then $\pi_1(M) \in \mathcal{C}$ and hence $\pi_1(M) \in \mathcal{A}_F$;*
(9) *If the group G possesses an upper bound on the orders of its finite subgroups and belongs to one of the following classes below, then G belongs to $\mathcal{A}_F$:*
   (a) *Residually {torsionfree elementary amenable} groups;*
   (b) *Free by elementary amenable groups;*
   (c) *Braid groups;*
   (d) *Right-angled Artin and Coxeter groups;*
   (e) *Torsion-free p-adic analytic pro-p-groups;*
   (f) *Locally indicable groups;*
   (g) *One-relator groups.*

PROOF.

(1) This is due to Linnell; see, for instance, [**77**] or [**87**, Theorem 10.19 on p. 378].
(2) This follows from [**55**, Corollary 1.2], which is based on on [**27**, Theorem 1.4].
(3) This follows from [**58**, Proposition 6.5].
(4) This follows from [**87**, Theorem 6.29 (2) on page 253].
(5) See [**87**, Lemma 10.4 on page 371].
(6) This follows from assertions (4) and (5).
(7) This follows from[**87**, Lemma 13.45 on page 473].
(8) This follows from [**64**, Theorem 1.1] for $d = 3$. The case $d = 2$ can be reduced to the case $d = 3$ by crossing with S^1 and assertion (4).
(9) This follows from other assertions or from [**55**, Theorem 1.1 and Corollary 1.2] using [**75**, Theorem 2] and [**33**, Theorem 1.1]. □

REMARK 3.6. The class $\mathcal{A}_F$ is very large by aforementioned results. Nevertheless, we do not know whether the Atiyah Conjecture 3.1 holds for all hyperbolic groups or for all amenable groups.

There are partial results on the difficult question, whether the Atiyah Conjecture 3.1 holds for a group G if it holds for a subgroup of finite index; see, for instance, [**76**].

REMARK 3.7. Dropping the condition on an upper bound on the orders of finite subgroups, one might still ask if the L^2-Betti numbers are always rational. This goes back to Atiyah's original question [**9**, p. 72], which asked for rationality of L^2-Betti numbers of closed manifolds. Austin [**10**, Corollary 1.2] gave the first example of a finitely generated group G, where for some matrix $A \in M_{m,n}(\mathbb{Q}G)$ the dimension $\dim_{\mathcal{N}(G)}(\ker(r_A))$ of the kernel of the $\mathcal{N}(G)$-homomorphism $r_A\colon \mathcal{N}(G)^m \to \mathcal{N}(G)^n$ is irrational. Grabowski [**43**, Theorem 1.3] proved, using Turing machines, that any nonnegative real number can occur in this way for some finitely generated group G and some matrix A. Löh and Uschold [**79**] investigate the computability degree of real numbers arising as L^2-Betti numbers or L^2-torsion of groups, parametrized over the Turing degree of the word problem. Roughly speaking, the complexity of the computation of L^2-invariants of a group is the same as the complexity of the word problem. This is due to the combinatorial computation in terms of characteristic sequences mentioned in Remark 9.1.

3.3. Embedding the group ring of a torsion-free group into a skew field. Associated to the von Neumann algebra $\mathcal{N}(G)$ is the algebra of affiliated operators $\mathcal{U}(G)$ which contains $\mathcal{N}(G)$. It can be defined analytically or just as the Ore localization of $\mathcal{N}(G)$ with respect to the multiplicative subset of nonzero divisors. Now one can consider the so-called division closure $\mathcal{D}(G)$ of $\mathbb{C}G$ in $\mathcal{U}(G)$, i.e., the smallest ring $\mathbb{C}G \subset \mathcal{D}(G) \subset \mathcal{U}(G)$ such that if $x \in \mathcal{D}(G)$ is invertible in $\mathcal{U}(G)$, its inverse is already contained in $\mathcal{D}(G)$.

The proof of the following is based on ideas of Peter Linnell from [**77**], which have been elaborated and expanded on in [**87**, Theorem 8.29 on p. 330 and Lemma 10.39 on p. 388] and [**110**]; see also [**36**, Theorem 3.8 (1) and 2]. The proofs given in the references above for $F = \mathbb{C}$ and $F = \mathbb{Q}$ carry directly over to arbitrary F.

THEOREM 3.8 (Main properties of $\mathcal{D}(G)$). *Let G be a torsion-free group. Consider a field F with $\mathbb{Q} \subseteq F \subseteq \mathbb{C}$.*

(1) *The group G satisfies the Atiyah Conjecture with coefficients in F if and only if $\mathcal{D}(G)$ is a skew field;*
(2) *Suppose that G satisfies the Atiyah Conjecture. Let C_* be a projective FG-chain complex. Then we get for all $n \geq 0$*

$$b_n^{(2)}\big(\mathcal{N}(G) \otimes_{FG} C_*\big) = \dim_{\mathcal{D}(G)}\big(H_n(\mathcal{D}(G) \otimes_{\mathbb{Q}G} C_*)\big).$$

In particular, $b_n^{(2)}\big(\mathcal{N}(G) \otimes_{FG} C_*\big)$ is an integer or ∞.

Theorem 3.8 shows that the Atiyah Conjecture 3.1 is related to the question whether for a torsion-free group G the group ring FG can be embedded into a skew field; see, for instance, [**51**]. Note that the existence of an embedding of FG into a skew field implies that FG has no nontrivial zero-divisors which is predicted by the Zero-Divisor Conjecture of Kaplansky.

3.4. The Algebraic Atiyah Conjecture. Recall that Linnell's program to approach the Atiyah Conjecture consists of a K-theoretic part and a ring theoretic part, namely that the composite

$$\operatorname{colim}_{H \subseteq G, |H| < \infty} K_0(FH) \to K_0(FG) \to K_0(\mathcal{D}(G))$$

is surjective and that $\mathcal{D}(G)$ is semisimple, see [**87**, Lemma 10.28 on p. 382 and Theorem 10.38 on p. 387]. There is the following so-called Algebraic Atiyah Conjecture, which is a purely K-theoretic statement and taken from [**56**, Conjecture 7.2], cf. [**87**, Lemma 10.26 on page 382]. It implies the Atiyah Conjecture 3.1; see [**87**, Lemma 10.26 on page 382].

CONJECTURE 3.9 (Algebraic Atiyah Conjecture). *Let G be a group and let F be a field F with $\mathbb{Q} \subseteq F \subseteq \mathbb{C}$, which is closed under complex conjugation. Let $\mathcal{R}_{FG}$ be the $*$-regular closure of FG in $\mathcal{U}(G)$.*

We say that G satisfies the Algebraic Atiyah Conjecture with coefficients in F *if the composite of the canonical maps*

$$\operatorname{colim}_{H \subseteq G, |H| < \infty} K_0(FH) \to K_0(FG) \to K_0(\mathcal{R}_{FG})$$

is surjective.

Recall that the Farrell–Jones Conjecture predicts the bijectivity of the map $\operatorname{colim}_{H\subseteq G,|H|<\infty} K_0(FH) \to K_0(FG)$. It is known for a large class of groups containing hyperbolic groups and CAT(0)-groups; see [**93**, Theorem 16.1]. If the Farrell–Jones Conjecture holds, then the Algebraic Atiyah Conjecture 3.9 is equivalent to the claim that the map $K_0(FG) \to K_0(\mathcal{R}_{FG})$ is surjective. For a group G, for which there is a bound on its finite subgroups, the Algebraic Atiyah Conjecture 3.9 is equivalent to the so-called center-valued Atiyah Conjecture with coefficients in F, which implies the Atiyah Conjecture 3.1 with coefficients in F; see [**50**, Definition 2.5.1, Proposition 2.5.2, Theorem 3.1.4] and also [**67**, Definition 1.2, Proposition 1.3].

For more information about the Atiyah Conjecture we refer, for instance, to [**87**, Chapter 10].

4. The Singer Conjecture

We now turn attention to a series of conjectures about L^2-invariants of aspherical manifolds. They deal with the phenomenon that L^2-invariants often vanish outside of the middle dimension.

4.1. Statement of the Singer Conjecture.

CONJECTURE 4.1 (Singer Conjecture). *If M is an aspherical closed topological manifold, then we get for $n \geq 0$*

$$b_n^{(2)}(\widetilde{M}) = 0 \quad \text{if } 2n \neq \dim(M).$$

If M is an aspherical closed topological manifold of even dimension $2m$, then

$$(-1)^m \cdot \chi(M) = b_m^{(2)}(\widetilde{M}) \geq 0.$$

If M is a closed connected smooth manifold of even dimension $2m$ admitting a Riemannian metric of negative sectional curvature, then

$$(-1)^m \cdot \chi(M) = b_m^{(2)}(\widetilde{M}) > 0.$$

The equality $(-1)^m \cdot \chi(M) = b_m^{(2)}(\widetilde{M})$ appearing in the Singer Conjecture 4.1 above follows from the the Euler–Poincaré formula $\chi(M) = \sum_{p\geq 0}(-1)^p \cdot b_p^{(2)}(\widetilde{M})$.

The Singer Conjecture 4.1 is consistent with the Atiyah Conjecture in the sense that it predicts that the L^2-Betti numbers $b_n^{(2)}(\widetilde{M})$ for an aspherical closed manifold M are all integers.

In original versions of the Singer Conjecture 4.1 the condition aspherical closed manifolds was replaced by the condition closed Riemannian manifold with nonpositive sectional curvature. Note that a closed Riemannian manifold with nonpositive sectional curvature is aspherical by Hadamard's Theorem.

4.2. Hopf Conjectures. Obviously the Singer Conjecture 4.1 implies the following conjecture in the cases, where M is aspherical or has negative sectional curvature.

CONJECTURE 4.2 (Hopf Conjecture). *If M is an aspherical closed topological manifold of even dimension $\dim(M) = 2m$, then*

$$(-1)^m \cdot \chi(M) \geq 0.$$

If M is a closed smooth manifold of even dimension $\dim(M) = 2m$ *with Riemannian metric and with sectional curvature* $\sec(M)$*, then*

$$\begin{array}{rclcrcl} (-1)^m \cdot \chi(M) & > & 0 & \textit{if} & \sec(M) & < & 0; \\ (-1)^m \cdot \chi(M) & \geq & 0 & \textit{if} & \sec(M) & \leq & 0; \\ \chi(M) & = & 0 & \textit{if} & \sec(M) & = & 0; \\ \chi(M) & \geq & 0 & \textit{if} & \sec(M) & \geq & 0; \\ \chi(M) & > & 0 & \textit{if} & \sec(M) & > & 0. \end{array}$$

The following version of the Hopf Conjecture for L^2-torsion appears in [**87**, Conjecture 11.3 on p. 418]

CONJECTURE 4.3 (Hopf Conjecture for L^2-torsion). *If M is an aspherical closed topological manifold of odd dimension* $\dim(M) = 2m+1$*, then M is* det-L^2*-acyclic, and its L^2-torsion satisfies*

$$(-1)^m \cdot \rho^{(2)}(\widetilde{M}) \geq 0.$$

If M is a closed smooth manifold of odd dimension $\dim(M) = 2m+1$ *with Riemannian metric and with sectional curvature* $\sec(M) < 0$*, then M is* det-L^2*-acyclic, and its L^2-torsion satisfies*

$$(-1)^m \cdot \rho^{(2)}(\widetilde{M}) > 0.$$

If M is an aspherical closed topological manifold of odd dimension $\dim(M) = 2m+1$ *whose fundamental group contains an amenable infinite normal subgroup, then M is* det-L^2*-acyclic and its L^2-torsion satisfies*

$$\rho^{(2)}(\widetilde{M}) = 0.$$

The Hopf Conjecture 4.3 for L^2-torsion is known to be true if one of the following conditions is satisfied:

(1) $\dim(M) \leq 3$;
(2) M is a locally symmetric space;
(3) $\pi_1(M)$ contains an elementary amenable infinite normal subgroup;
(4) M carries a nontrivial S^1-action.

Statement (1) follows from combining [**97**, Theorem 0.7] with Thurston's Geometrization Conjecture, which is known to be true by [**66**, **101**], following the spectacular outline of Perelman. Statement (2) can be found in [**87**, Corollary 5.16 on page 231] and statements (3) and (4) are from [**89**, Corollary 1.13]. See also Remark 6.12.

4.3. Status of the Singer Conjecture. The Singer Conjecture 4.1 is known for an aspherical closed smooth manifold M in the following cases:

(1) $\dim(M) \leq 3$;
(2) M comes with a Riemannian metric whose sectional curvature is negative and satisfies certain pinching conditions;
(3) M is a locally symmetric space;
(4) M possesses a Riemannian metric, whose sectional curvature is negative and M carries some Kähler structure;
(5) M is an aspherical closed Kähler manifold, whose fundamental group is word-hyperbolic in the sense of Gromov [**46**].
(6) $\pi_1(M)$ contains an amenable infinite normal subgroup.
(7) M carries a nontrivial S^1-action;
(8) M fibers over S^1.

The precise statements and proofs and the relevant references, e.g., [**15, 24, 47, 59, 83, 103**], can be found in [**87**, Theorem 1.39 on p. 42, Corollary 1.43 on p. 48, Theorem 1.44 on p. 48, Section 11.1], again using Thurston's Geometrization Conjecture for the three dimensional case. We also mention that proofs of (1)–(5) use geometric properties of the manifold whereas (6)–(8) use algebraic topological techniques and do not use the manifold structure.

Partial results about the Singer Conjecture 4.1 for right-angled Coxeter groups can be found in Davis and Okun [**26**].

The paper by Albanese, Di Cerbo, and Lombardi [**6**] deals with the Singer Conjecture for aspherical complex surfaces and proves it for aspherical complex surfaces with residually finite fundamental groups in [**6**, Theorem 1.5].

In contrast to the Atiyah Conjecture, evidence for the Singer Conjecture 4.1 comes from computations only and no promising proof strategy is known. In some sense, Poincaré duality together with L^2-bounds on differential forms on $\widetilde{M}$ seem to force the L^2-Betti numbers $b_p^{(2)}(\widetilde{M})$ of an aspherical closed manifold to concentrate in the middle dimension. One may wonder what happens if we replace M by an aspherical finite Poincaré complex in the Singer Conjecture 4.1. There are counterexamples to the Singer Conjecture 4.1 if one weakens aspherical to rationally aspherical, see [**11**, Theorem 4]. The reader should also take a look at Remarks 6.10 and 6.12.

4.4. The proper Singer Conjecture and the action dimension. One may generalize the Singer Conjecture 4.1 to the following conjecture.

CONJECTURE 4.4 (Proper Singer Conjecture). *A group G satisfies the* proper Singer Conjecture, *if for any contractible topological manifold M without boundary on which G acts properly and cocompactly, we have for all $n \geq 0$*

$$b_n^{(2)}(M; \mathcal{N}(G)) = 0 \quad \textit{if } 2n \neq \dim(M).$$

Note that the proper Singer Conjecture is a statement about a group G, whereas the Singer Conjecture 4.1 is a statement about an aspherical closed manifold. Provided that G is torsion-free, the proper Singer Conjecture holds for G if and only if the Singer Conjecture 4.1 holds for one (and hence all) aspherical closed manifold M with $\pi_1(M) \cong G$.

The proper Singer Conjecture for a group G is equivalent to the cadim conjecture (cadim standing for compact action dimension) for G for manifolds with PL-boundary; see [**102**, Theorem 4.10]. The latter predicts that for a contractible topological manifold M (possibly with boundary) that admits a cocompact proper topological action of G such that the boundary ∂M of M carries a PL-structure for which the G-action on ∂M is through PL-automorphisms we have

$$2 \cdot \sup\{n \in \mathbb{Z}^{\geq 0} \mid b_n^{(2)}(G) \neq 0\} \leq \dim(M).$$

For more informations about the notion of the (compact) action dimension and its relationship to L^2-Betti numbers we refer for instance to [**12, 18, 102**].

4.5. The $\mathbb{F}_p$-Singer Conjecture. One may also consider the *$\mathbb{F}_p$-Singer Conjecture*, first stated in [**13**], which predicts that for an aspherical closed topological manifold M and any prime p we get for $n \geq 0$

$$b_n^{(2)}(\pi_1(M); \mathbb{F}_p) = 0 \quad \text{if } 2n \neq \dim(M).$$

Here $b_n^{(2)}(\pi_1(M);\mathbb{F}_p)$ is defined only if $\pi_1(M)$ is residually finite. Namely, for a chain of normal subgroups of finite index $\pi_1(M) = \Gamma_0 \supseteq \Gamma_1 \supseteq \Gamma_2 \supseteq \cdots$ with $\bigcap_{n=0}^{\infty} \Gamma_n = \{1\}$, one puts $b_n^{(2)}(\pi_1(M);\mathbb{F}_p) = \limsup_{n\to\infty} \frac{b_n(\Gamma_k;\mathbb{F}_p)}{[\Gamma:\Gamma_k]}$. It is unknown in general whether this definition depends on the chain of subgroups Γ_k and the statement above is to be understood in the sense that it holds for any such chain. The definition of L^2-Betti numbers over $\mathbb{F}_p$ is motivated by the Approximation Theorem in characteristic zero 2.4(14). The $\mathbb{F}_p$-Singer Conjecture is open for $\dim(M) = 3$. However, for any odd prime p, the $\mathbb{F}_p$-Singer Conjecture fails in all odd dimensions larger or equal to 7 and all even dimensions larger or equal to 14; see [**13**, Theorem 4].

5. The Determinant Conjecture

Next we explain the Determinant Conjecture [**87**, Conjecture 13.2 on p. 454]. It is needed for the definition of L^2-determinant class and implies homotopy invariance of L^2-torsion; see Theorem 2.4(1). Furthermore, it implies the Approximation Conjecture; see Remark 7.2.

CONJECTURE 5.1 (Determinant Conjecture for a group G). *For any matrix $A \in M_{r,s}(\mathbb{Z}G)$, the Fuglede–Kadison determinant of the $\mathcal{N}(G)$-homomorphism $r_A\colon \mathcal{N}(G)^r \to \mathcal{N}(G)^s$ given by right multiplication with A satisfies*

$$\det_{\mathcal{N}(G)}(r_A) \geq 1.$$

The class $\mathcal{F}$ of groups for which it is true has the following properties:

(1) *Trivial group.*
The trivial group belongs to $\mathcal{F}$;
(2) *Amenable quotient.*
Let $H \subset G$ be a normal subgroup. Suppose that $H \in \mathcal{F}$ and the quotient G/H is amenable. Then $G \in \mathcal{F}$;
(3) *Filtered colimits.*
The class $\mathcal{F}$ is closed under countable filtered colimits;
(4) *Cofiltered limits.*
The class $\mathcal{F}$ is closed under countable cofiltered limits;
(5) *Subgroups.*
If H is isomorphic to a subgroup of a group G with $G \in \mathcal{F}$, then $H \in \mathcal{F}$.
(6) *Quotients with finite kernel.*
Let $1 \to K \to G \to Q \to 1$ be an exact sequence of groups. If K is finite and G belongs to $\mathcal{F}$, then Q belongs to $\mathcal{F}$;
(7) *Sofic groups.*
Sofic groups belong to $\mathcal{F}$.

For the verification of the Determinant Conjecture for the groups above, see [**29**, Theorem 5], [**87**, Section 13.2 on pp. 459 ff], [**116**, Theorem 1.21].

To sketch how large $\mathcal{F}$ really is, let us mention that already the class of sofic groups is very large. It is closed under direct and free products, taking subgroups, taking inverse and direct limits over directed index sets, and is closed under extensions with amenable groups as quotients and a sofic group as kernel. In particular, it contains all residually amenable groups. One expects that there exists nonsofic groups but no example is known. More information about sofic groups can be found, for instance, in [**30**] and [**105**]. More information about the Determinant Conjecture 5.1 can be found in [**87**, Chapter 13] and [**116**].

6. Approximation for finite index normal chains

Recall that the approximation (see Theorem 2.4(14)) predicts that L^2-Betti numbers of spaces can be computed as limits of normalized ordinary Betti numbers of certain towers of coverings. In this section we discuss this phenomenon for more general invariants.

6.1. Basic setup for approximation for finite index normal chains. Let G be a (discrete) group. A *finite index normal chain* $\{G_i\}$ for G is a descending chain of subgroups

$$G = G_0 \supseteq G_1 \supseteq G_2 \supseteq \cdots \tag{6.1}$$

such that G_i is normal in G, the index $[G : G_i]$ is finite and $\bigcap_{i \geq 0} G_i = \{1\}$.

Let $p\colon \overline{X} \to X$ be a G-covering. Put $X[i] := G_i \backslash \overline{X}$. We obtain a $[G : G_i]$-sheeted covering $p[i]\colon X[i] \to X$. Its total space $X[i]$ inherits the structure of a finite CW-complex, a closed manifold, or a closed Riemannian manifold respectively if X has the structure of a finite CW-complex, a closed manifold, or a closed Riemannian manifold respectively.

Let α be a classical topological invariant such as the Euler characteristic, the signature, the nth Betti number with coefficients in the field $\mathbb{Q}$ or $\mathbb{F}_p$, torsion in the sense of Reidemeister or Ray and Singer, or the logarithm of the cardinality of the torsion subgroup of the nth homology group with integral coefficients. We want to study the sequence

$$\left(\frac{\alpha(X[i])}{[G : G_i]}\right)_{i \geq 0}.$$

PROBLEM 6.1 (Approximation Problem).

(1) *Does the sequence converge?*
(2) *If yes, is the limit independent of the chain $\{G_i\}$?*
(3) *If yes, what is the limit?*

The hope is that the answer to the first two questions is yes and the limit turns out to be an L^2-analog $\alpha^{(2)}$ of α applied to the G-space $\overline{X}$, i.e., one can prove an equality of the type

$$\lim_{i \to \infty} \frac{\alpha(X[i])}{[G : G_i]} = \alpha^{(2)}(\overline{X}; G). \tag{6.2}$$

Equation (6.2) is often used to compute the L^2-invariant $\alpha^{(2)}(\overline{X}; G)$ by its finite-dimensional analogues $\alpha(X[i])$. On the other hand, it implies the existence of finite coverings with large $\alpha(X[i])$, if $\alpha^{(2)}(\overline{X}; G)$ is known to be positive.

6.2. The Euler characteristic. The *Euler characteristic* $\chi(X)$ of a finite CW-complex is multiplicative under finite coverings. Because this implies $\chi(X) = \frac{\chi(X[i])}{[G:G_i]}$, the answer in this case is yes to the questions appearing in Problem 6.1, and the limit is

$$\lim_{i \to \infty} \frac{\chi(X[i])}{[G : G_i]} = \chi(X). \tag{6.3}$$

6.3. The signature. Next we consider the *signature* $\operatorname{sign}(M)$ of a closed oriented topological $4k$-dimensional manifold M. It is known that it is multiplicative under finite coverings; however, the proof is more involved than the one for the Euler characteristic. It follows, for instance, from Hirzebruch's Signature Theorem (see [**53**]) or Atiyah's L^2-index theorem [**9**, (1.1)] in the smooth case; for closed topological manifolds see Schafer [**115**, Theorem 8]. Since this implies $\operatorname{sign}(X) = \frac{\operatorname{sign}(X[i])}{[G:G_i]}$, each of the questions appearing in Problem 6.1 has a positive answer, and the limit is

$$\lim_{i\to\infty} \frac{\operatorname{sign}(X[i])}{[G:G_i]} = \operatorname{sign}(X). \tag{6.4}$$

The next level of generality is to pass from a oriented closed topological manifold to a oriented finite Poincaré complex, whose definition is due to Wall [**125**]. For them the signature is still defined if the dimension is divisible by 4. There are Poincaré complexes X for which the signature is not multiplicative under finite coverings; see [**108**, Example 22.28], [**125**, Corollary 5.4.1]. Hence the situation is more complicated here. Nevertheless, it turns out in this case each of the questions appearing in Problem 6.1 has a positive answer, and the limit is

$$\lim_{i\to\infty} \frac{\operatorname{sign}(X[i])}{[G:G_i]} = \operatorname{sign}^{(2)}(\overline{X};\mathcal{N}(G)), \tag{6.5}$$

where $\operatorname{sign}^{(2)}(\overline{X};\mathcal{N}(G))$ denotes the L^2-signature, which is in general different from $\operatorname{sign}(X)$ for a finite Poincaré complex X.

For more information and details we refer to [**98, 99**].

6.4. Approximation of L^2-Betti numbers in characteristic zero. Fix a field F of characteristic zero. We consider the *nth Betti number with F-coefficients* $b_n(X;F) := \dim_F(H_n(X;F))$. Note that $b_n(X;F) = b_n(X;\mathbb{Q}) = \operatorname{rk}_{\mathbb{Z}}(H_n(X;\mathbb{Z}))$ holds, where $\operatorname{rk}_{\mathbb{Z}}$ denotes the rank of a finitely generated abelian group. In this case each of the questions appearing in Problem 6.1 has a positive answer by the main result of Lück's article [**82**].

THEOREM 6.2. *Let F be a field of characteristic zero, and let X be a finite CW-complex. Then for each finite index normal chain $\{G_i\}$ we have*

$$\lim_{i\to\infty} \frac{b_n(X[i];F)}{[G:G_i]} = b_n^{(2)}(\overline{X};\mathcal{N}(G)),$$

where $b^{(2)}(\overline{X};\mathcal{N}(G))$ denotes the nth L^2-Betti number.

Löh and Uschold [**79**, Proposition 6.6] prove a quantitative version of Theorem 6.2. Essentially they show

$$\left| b_n^{(2)}(\overline{X};\mathcal{N}(G)) - \frac{b_n(X[i];F)}{[G:G_i]} \right| \leq a\cdot\left(1-\frac{1}{b\cdot i}\right)^{i^2} + \frac{a\cdot\log(b)}{\log(i)}$$

for certain constants a and b depending on X.

6.5. Approximation of L^2-Betti numbers in prime characteristic. The situation is more complicated and unclear in prime characteristic. Fix a prime p. Let F be a field of characteristic p. We consider the *nth Betti number with F-coefficients* $b_n(X;F) := \dim_F(H_n(X;F))$. Note that $b_n(X;F) = b_n(X;\mathbb{F}_p)$ holds where $\mathbb{F}_p$ is the field of p-elements. In this setting a general answer to Problem 6.1

is only known in special cases. The main problem is that one does not have an analogue of the von Neumann algebra in characteristic p and the construction of an appropriate extended dimension function, see [**85**], is not known in general.

If G is torsion-free elementary amenable, one gets the full positive answer by Linnell-Lück-Sauer [**74**, Theorem 0.2], where more explanations, e.g., about Ore localizations, are given and actually virtually torsion-free elementary amenable groups are considered.

THEOREM 6.3. *Let F be a field (of arbitrary characteristic) and X be a connected finite CW-complex. Let G be a torsion-free elementary amenable group. Then for each finite index normal chain $\{G_i\}$:*

$$\dim_{FG}^{\mathrm{Ore}}(H_n(\overline{X};F)) = \lim_{n\to\infty} \frac{b_n(X[i];F)}{[G:G_n]}.$$

Note that Theorem 6.3 is consistent with Theorem 6.2 since for a field F of characteristic zero and a torsion-free elementary amenable group G we have $b_n^{(2)}(\overline{X};\mathcal{N}(G)) = \dim_{FG}^{\mathrm{Ore}}(H_n(\overline{X};F))$. The latter equality follows from [**87**, Theorem 6.37 on p. 259, Theorem 8.29 on p. 330, Lemma 10.16 on p. 376, and Lemma 10.39 on p. 388].

Here is another special case taken from Bergeron, Lück, Linnell, and Sauer [**16**], (see also Calegari and Emerton [**22, 23**]), where we know the answer only for special chains. Let p be a prime, let n be a positive integer, and let $\phi\colon G \to \mathrm{GL}_n(\mathbb{Z}_p)$ be a homomorphism, where $\mathbb{Z}_p$ denotes the p-adic integers. The closure of the image of ϕ, which is denoted by Λ, is a p-adic analytic group admitting an exhausting filtration by open normal subgroups $\Lambda_i = \ker\big(\Lambda \to \mathrm{GL}_n(\mathbb{Z}/p^i\mathbb{Z})\big)$. Put $G_i = \phi^{-1}(\Lambda_i)$.

THEOREM 6.4. *Let F be a field (of arbitrary characteristic). Put $d = \dim(\Lambda)$. Let X be a finite CW-complex. Then for any integer n and as i tends to infinity, we have:*

$$b_n(X[i];F) = b_n^{(2)}(\overline{X};F)\cdot[G:G_i] + O\left([G:G_i]^{1-1/d}\right),$$

where $b_n^{(2)}(\overline{X};F)$ is the nth mod p L^2-Betti numbers occurring in [**16**, Definition 1.3] *which is defined using homology coefficients in the Iwasawa algebra of G. In particular,*

$$\lim_{i\to\infty} \frac{b_n(X[i];F)}{[G:G_i]} = b_n^{(2)}(\overline{X};F).$$

Returning to the setting of arbitrary finite index normal chain $(G_i)_{i\geq 0}$, we get by the universal coefficient theorem $b_n(X[i];\mathbb{Q}) \leq b_n(X[i];F)$ for any field F and hence by Theorem 6.2 the inequality

$$\liminf_{i\to\infty} \frac{b_n(X[i];F)}{[G:G_i]} \geq b_n^{(2)}(\overline{X};\mathcal{N}(G)).$$

If p is a prime and we also assume that each index $[G:G_i]$ is a p-power, then the sequence $\frac{b_n(X[i];F)}{[G:G_i]}$ is monotone decreasing and, in particular, $\lim_{i\to\infty} \frac{b_n(X[i];F)}{[G:G_i]}$ exists; see [**16**, Theorem 1.6].

QUESTION 6.5 (Approximation in prime characteristic). *Does the sequence $\frac{b_n(X[i];F)}{[G:G_i]}$ converge and is the limit $\lim_{i\to\infty} \frac{b_n(X[i];F)}{[G:G_i]}$ independent of the chain $\{G_i\}$ for all fields F and $n \geq 0$, provided that X is finite and $\overline{X}$ is contractible?*

REMARK 6.6. The third author conjectured in the situation of Question 6.5 that the $\lim_{i\to\infty} \frac{b_n(X[i];F)}{[G:G_i]}$ is equal to $b_n^{(2)}(X;\mathcal{N}(G))$ for every field F and for all finite index normal chains $\{G_i\}$, see [**90**, Conjecture 3.4 on p. 275]. This is true in characteristic zero by Theorem 6.2 and if G is torsion-free elementary amenable by Theorem 6.3, but not in prime characteristic by Avramidi, Okun, and Schreve [**13**, Corollary 2]. Namely, Avramidi, Okun, and Schreve [**13**, Theorem 1] prove that for a right-angled Artin group A_L with defining flag complex L and any field F the limit $\lim_{i\to\infty} \frac{b_n(X[i];F)}{[G:G_i]}$ exists, is independent of the chain $\{G_i\}$, and actually agrees with the reduced Betti number $\overline{b}_{n-1}(L;F)$ of L with coefficients in F. Furthermore, they construct examples where $\overline{b}_3(L;\mathbb{Q}) = 0$ and $\overline{b}_3(L;\mathbb{F}_2) = 1$ hold. These counterexamples do not exist in degree $n = 1$.

REMARK 6.7 (L^2-Betti numbers in finite characteristic via skew fields). Recall from Theorem 3.8 that if a torsion-free group G satisfies the Atiyah Conjecture, then the group ring $\mathbb{C}G$ embeds into a skew field $\mathcal{D}(G)$ and L^2-Betti numbers are the Betti numbers with coefficients in this skew field. Motivated by this, one might hope that for a torsion-free group G and any field F there exists a skew field $\mathcal{D}_F(G)$ together with an embedding $FG \hookrightarrow \mathcal{D}_F(G)$, which can be used to define L^2-Betti numbers with coefficients in F as $b_n^{(2)}(\overline{X};F) := \dim_{\mathcal{D}_F(G)} H_n^G(\overline{X};\mathcal{D}_F(G))$. An approach to Question 6.5 would then be to show that the sequence

$$\frac{b_n(X[i];F)}{[G:G_i]}$$

converges to $b_n^{(2)}(\overline{X};F)$ for $i \to \infty$.

Jaikin-Zapirain constructs such embeddings in [**57**, Corollary 1.3] for large classes of groups. He shows that such an embedding of FG into a skew field with very nice properties (it is a Hughes free division ring and the universal division ring of fractions of FG) exists if G is residually (torsion-free nilpotent) or residually (locally indicable amenable). For a residually (torsion-free nilpotent) group G Avramidi, Okun, and Schreve show in [**14**, Theorem 3.6] the following version of the approximation conjecture

$$b_n^{(2)}(\overline{X};F) = \inf\left\{\frac{b_n(\overline{X}/H;F)}{[G:H]} : H \leq G, [G:H] < \infty\right\}. \tag{6.6}$$

This equality also holds if G is residually (locally indicable admitting a finite index normal chain) as observed in [**14**, Remarks on p. 326]. In the case $F = \mathbb{Q}$ of the classical Approximation Theorem 6.2 the arguably easier Kazhdan's inequality predicts that L^2-Betti numbers are bounded by

$$b_n^{(2)}(\overline{X};\mathcal{N}(G)) \geq \limsup_{i\to\infty} \frac{b_n(X[i];\mathbb{Q})}{[G:G_i]}$$

for any finite index normal chain $\{G_i\}$ for G. It is not clear, however, whether an analog of this estimate holds in the present setting. This would imply that (6.6) still holds when one replaces inf by the limit over a finite index normal chain.

6.6. Torsion. For a detailed discussion of approximation for finite index normal chains for torsion invariants such as the Ray-Singer torsion or the integral torsion we refer to [**90**, Sections 8–10] and also to Section 7.3.

6.7. Homological torsion growth and L^2-torsion. The following conjecture is taken from [**89**, Conjecture 1.12 (2)]. For locally symmetric spaces it reduces to the conjecture of Bergeron and Venkatesh [**17**, Conjecture 1.3].

CONJECTURE 6.8 (Homological torsion growth and L^2-torsion). *Let M be an aspherical closed manifold.*

Then we get for any natural number n with $2n+1 \neq \dim(M)$

$$\lim_{i\to\infty} \frac{\ln\left(\left|\operatorname{tors}\left(H_n(M[i];\mathbb{Z})\right)\right|\right)}{[G:G_i]} = 0.$$

If the dimension $\dim(M) = 2m+1$ is odd, then $\widetilde{M}$ is det-*L^2-acyclic and we get*

$$\lim_{i\to\infty} \frac{\ln\left(\left|\operatorname{tors}\left(H_m(M[i];\mathbb{Z})\right)\right|\right)}{[G:G_i]} = (-1)^m \cdot \rho^{(2)}(\widetilde{M}).$$

Conjecture 6.8 is true if $\pi_1(M)$ contains an elementary amenable infinite normal subgroup or M carries a nontrivial S^1-action; see [**89**, Corollary 1.13].

REMARK 6.9 (Criteria for vanishing homology torsion growth). Recently, there has been some progress in proving vanishing of homology growth and torsion homology growth for certain groups. These results are based on the general vanishing result of Abert, Bergeron, Fraczyk, and Gaboriau [**2**, Theorem 10.20] and have been applied to show vanishing for special linear groups and mapping class groups [**2**, Theorem A, D]. Further applications of this theorem can be found in [**7**, Theorem A], [**8**, Theorem A], and [**124**, Proposition 1.6], where vanishing of homology growth and torsion homology growth of mapping tori of finitely generated free groups with respect to a polynomially growing automorphism and of right-angled Artin groups which are inner amenable is shown. Combining the first result with [**25**, Theorem 5.1] confirms Conjecture 6.8 for mapping tori of aspherical closed manifolds which have finitely generated free fundamental group and the monodromy is polynomially growing.

REMARK 6.10 (The Singer Conjecture 4.1 and Conjecture 6.8 are not compatible). The Singer Conjecture 4.1 and Conjecture 6.8 about homological torsion growth and L^2-torsion cannot both be true in general. Namely, if both are true, then the $\mathbb{F}_p$-Singer Conjecture of Section 4.5 would be true as pointed out by Avramidi, Okun, and Schreve before Theorem 4 appearing in [**13**]. The argument is as follows.

Suppose we have an aspherical closed d-dimensional manifold M and both the Singer Conjecture 4.1 and Conjecture 6.8 are true. Consider n with $2n \neq d$. We have to show

$$\lim_{i\to\infty} \frac{b_n(M[i];\mathbb{F}_p)}{[G:G_i]} = 0 \tag{6.7}$$

for $G = \pi_1(M)$ and a finite index normal chain $\{G_i\}$. Because of Poincaré duality it suffices to consider the case $2n > d$. Put $M^3 = M \times M \times M$, $G^3 = G \times G \times G$ and $G_i^3 = G_i \times G_i \times G_i$. Then $\widetilde{M^3} = \widetilde{M} \times \widetilde{M} \times \widetilde{M}$, $G^3 = \pi_1(M^3)$, and $M^3[i] = M[i]^3$. By the Singer Conjecture 4.1 applied to M^3 we get $b_{3n}^{(2)}(\widetilde{M^3}) = 0$. This implies by Theorem 6.3

$$\lim_{i\to\infty} \frac{\operatorname{rk}_{\mathbb{Z}}(H_{3n}(M[i]^3;\mathbb{Z}))}{[G^3:G_i^3]} = \lim_{i\to\infty} \frac{b_{3n}(M[i]^3;\mathbb{Q})}{[G^3:G_i^3]} = b_{3n}^{(2)}(\widetilde{M^3};\mathbb{Q}) = 0.$$

Since $2(3n) > 2(3n-1) > 3d$ holds, we conclude from Conjecture 6.8

$$\lim_{i\to\infty} \frac{\ln\left(\left|\operatorname{tors}\left(H_{3n}(M^3[i];\mathbb{Z})\right)\right|\right)}{[G^3 : G_i^3]} = 0,$$

$$\lim_{i\to\infty} \frac{\ln\left(\left|\operatorname{tors}\left(H_{3n-1}(M^3[i];\mathbb{Z})\right)\right|\right)}{[G^3 : G_i^3]} = 0.$$

We have for any $k \geq 0$

$$\begin{aligned}
\dim_{\mathbb{F}_p}(\mathbb{Z}^k \otimes_{\mathbb{Z}} \mathbb{F}_p) &= \dim_{\mathbb{Z}}(\mathbb{Z}^k);\\
\dim_{\mathbb{F}_p}(\mathbb{Z}/p^k \otimes_{\mathbb{Z}} \mathbb{F}_p) &= 1;\\
\dim_{\mathbb{F}_p}\left(\operatorname{Tor}_1^{\mathbb{Z}}(\mathbb{Z}/p^k, \mathbb{F}_p)\right) &= 1;\\
\ln\left(|\operatorname{tors}(\mathbb{Z}/p^k)|\right) &= k \cdot \ln(p).
\end{aligned}$$

If q is an prime different from p, we get $\mathbb{Z}/q^k \otimes_{\mathbb{Z}} \mathbb{F}_p = 0$ and $\operatorname{Tor}_1^{\mathbb{Z}}(\mathbb{Z}/q^k, \mathbb{F}_p) = 0$. Hence we get for any finitely generated abelian group M

$$\dim_{\mathbb{F}_p}(M \otimes_{\mathbb{Z}} \mathbb{F}_p) \leq \dim_{\mathbb{Z}}(M) + \frac{\ln\left(|\operatorname{tors}(M)|\right)}{\ln(2)};$$

$$\dim_{\mathbb{F}_p}\left(\operatorname{Tor}_1^{\mathbb{Z}}(M, \mathbb{F}_p)\right) \leq \frac{\ln\left(|\operatorname{tors}(M)|\right)}{\ln(2)}.$$

We conclude from the Künneth Formula and the Universal Coefficient Theorem

$$\begin{aligned}
\frac{b_n(M[i];\mathbb{F}_p)^3}{[G:G_i]^3} &= \frac{b_n(M[i];\mathbb{F}_p)^3}{[G^3:G_i^3]}\\
&\leq \frac{b_{3n}(M[i]^3;\mathbb{F}_p)}{[G^3:G_i^3]}\\
&= \frac{\dim_{\mathbb{F}_p}\left(H_{3n}(M[i]^3;\mathbb{Z}) \otimes_{\mathbb{Z}} \mathbb{F}_p\right) + \dim_{\mathbb{F}_p}\left(\operatorname{Tor}_1^{\mathbb{Z}}(H_{3n-1}(M[i];\mathbb{Z}), \mathbb{F}_p)\right)}{[G^3:G_i^3]}\\
&\leq \frac{\dim_{\mathbb{Z}}(H_{3n}(M[i]^3;\mathbb{Z}))}{[G^3:G_i^3]} + \frac{\ln\left(\left|\operatorname{tors}\left(H_{3n}(M^3[i];\mathbb{Z})\right)\right|\right)}{[G^3:G_i^3]\cdot\ln(2)}\\
&\qquad + \frac{\ln\left(\left|\operatorname{tors}\left(H_{3n-1}(M^3[i];\mathbb{Z})\right)\right|\right)}{[G^3:G_i^3]\cdot\ln(2)}.
\end{aligned}$$

Hence (6.7) is true.

We have already mentioned in Section 4.5 that the $\mathbb{F}_p$-Singer Conjecture is not true in general; see [**13**, Theorem 4]. Hence the Singer Conjecture 4.1 or Conjecture 6.8 is not true.

In view of Remark 6.10 we suggest keeping the Singer Conjecture 4.1 and weaken Conjecture 6.8 to the following version.

CONJECTURE 6.11 (Modified Homological torsion growth and L^2-torsion). *Let M be an aspherical closed manifold of odd dimension* $\dim(M) = 2m+1$.

Then M is det-L^2-*acyclic, the limit*

$$\lim_{i\to\infty} \left(\sum_{n=0}^{2m+1} (-1)^n \cdot \frac{\ln\left(|\operatorname{tors}(H_n(M[i];\mathbb{Z}))|\right)}{[G:G_i]}\right)$$

exists and is given by

$$\lim_{i\to\infty}\left(\sum_{n=0}^{2m+1}(-1)^n\cdot\frac{\ln\left(\left|\operatorname{tors}\left(H_n(M[i];\mathbb{Z})\right)\right|\right)}{[G:G_i]}\right)=\rho^{(2)}(\widetilde{M}).$$

REMARK 6.12 (Discussion of Conjecture 6.11). It is conceivable that both the Singer Conjecture 4.1 and the modified Conjecture 6.11 about homological torsion growth and L^2-torsion are true, since the argument in Remark 6.10 that the Singer Conjecture 4.1 or Conjecture 6.8 is not true does not apply anymore.

The difference between Conjecture 6.8 and Conjecture 6.11 is that we drop the claim $\lim_{i\to\infty}\frac{\ln(|\operatorname{tors}(H_n(M[i];\mathbb{Z}))|)}{[G:G_i]}=0$ in Conjecture 6.8 and allow in Conjecture 6.11 contributions from all $n\geq 0$ and not only from $n=m$ as in Conjecture 6.8. We conclude $\operatorname{tors}\big(H_n(M[i];\mathbb{Z})\big)\cong\operatorname{tors}\big(H_{2m-n}(M[i];\mathbb{Z})\big)$ for $n\geq 0$ from Poincaré duality and the Universal Coefficient Theorem and the vanishing of $\operatorname{tors}\big(H_0(M[i];\mathbb{Z})\big)$. We get the equality

$$\begin{aligned}&\sum_{n=0}^{2m+1}(-1)^n\cdot\frac{\ln\left(\left|\operatorname{tors}\left(H_n(M[i];\mathbb{Z})\right)\right|\right)}{[G:G_i]}\\&=(-1)^m\cdot\frac{\ln\left(\left|\operatorname{tors}\left(H_m(M[i];\mathbb{Z})\right)\right|\right)}{[G:G_i]}+2\cdot\sum_{n=1}^{m-1}(-1)^n\cdot\frac{\ln\left(\left|\operatorname{tors}\left(H_n(M[i];\mathbb{Z})\right)\right|\right)}{[G:G_i]}.\end{aligned}$$

Hence in dimension $\dim(M)=3$ Conjecture 6.8 and Conjecture 6.11 agree. This does not lead to any contradiction since no 3-dimensional counterexample to the $\mathbb{F}_p$-Singer Conjecture is known.

Note that Conjecture 6.11 does not imply the Hopf Conjecture for L^2-torsion 4.3 for aspherical closed topological manifolds. One may wonder whether the Hopf Conjecture for L^2-torsion 4.3 is not true in odd dimensions $\dim(M)\geq 5$ in general. It is known to be true in dimension $\dim(M)=3$ by [**97**, Theorem 0.7] yet again, since Thurston's Geometrization Conjecture is known to be true. The modified Conjecture 6.11 resolves the problem mentioned above, but may only be true in special cases.

One may replace in Conjecture 6.11 the aspherical closed manifold M of odd dimension $\dim(M)=2m+1$ by a connected finite CW-complex X, for which $\widetilde{X}$ is det-L^2-acyclic. This corresponds to [**90**, Conjecture 8.9 on page 290]. This version is true if $\pi_1(X)\cong\mathbb{Z}$ holds (see [**17**, Theorem 7.3]) but at the time of writing nothing is known in general if $\pi_1(X)$ is infinite and not equal to $\mathbb{Z}$.

For applications of these approximation conjectures for L^2-torsion to questions about profinite rigidity for fundamental groups of 3-manifolds and lattices in higher rank Lie groups we refer to [**60**, Section 6.7] and [**61**]. Estimates for the homological growth in terms of the volume for aspherical manifolds are established in [**113**].

6.8. Further examples. Other invariants, such as the rank gradient and the cost, truncated Euler characteristics, and minimal numbers of generators, are discussed in [**90**, Sections 4 and 5]. For considerations concerning the speed of convergence we refer to [**90**, Section 6].

7. Approximation for normal chains

A *normal chain* $\{G_i\}$ for the group G is a descending chain of subgroups

$$G = G_0 \supseteq G_1 \supseteq G_2 \supseteq \cdots \tag{7.1}$$

such that G_i is normal in G and $\bigcap_{i\geq 0} G_i = \{1\}$. Note that a normal chain is a finite index normal chain, if and only if $[G : G_i]$ is finite for each i. Next we want to discuss approximation results for normal chains.

7.1. The Approximation Conjecture for L^2-Betti numbers. Next we deal with the Approximation Conjecture for L^2-Betti numbers (see [**116**, Conjecture 1.10], [**87**, Conjecture 13.1 on p. 453]).

CONJECTURE 7.1 (Approximation Conjecture for L^2-Betti numbers). *A group G satisfies the* Approximation Conjecture for L^2-Betti numbers *if for any normal chain $\{G_i\}$ one of the following equivalent conditions holds*

(1) *Matrix version.*
Let $A \in M_{r,s}(\mathbb{Q}G)$ be a matrix. Then

$$\begin{aligned}\dim_{\mathcal{N}(G)}\bigl(\ker\bigl(r_A^{(2)}\colon L^2(G)^r \to L^2(G)^s\bigr)\bigr)\\ = \lim_{i\to\infty} \dim_{\mathcal{N}(G/G_i)}\bigl(\ker\bigl(r_{A[i]}^{(2)}\colon L^2(G/G_i)^r \to L^2(G/G_i)^s\bigr)\bigr);\end{aligned}$$

(2) *CW-complex version.*
Let X be a G-CW-complex of finite type. Then $X[i] := G_i\backslash X$ is a G/G_i-CW-complex of finite type and we get for $n \geq 0$

$$b_n^{(2)}(X;\mathcal{N}(G)) = \lim_{i\to\infty} b_n^{(2)}(X[i];\mathcal{N}(G/G_i)).$$

The two conditions appearing in Conjecture 7.1 are equivalent by [**87**, Lemma 13.4 on p. 455].

REMARK 7.2 (The Determinant Conjecture and the Approximation Conjecture). Let G be a group and $\{G_i\}$ be a normal chain. Suppose that G and each G/G_i satisfies the Determinant Conjecture 5.1. Then the Approximation Conjecture 7.1 for L^2-Betti numbers holds for G for this normal chain $\{G_i\}$ by [**87**, Theorem 13.3(1) on p. 454]. We mention that in [**87**, Theorem 13.3(1) on p. 454] there is a misprint—G_i has to be replaced by G/G_i. Moreover, as pointed out to us by Bin Sun, [**87**, Theorem 13.3(2) on p. 454] is not correct as stated. It is true that G is of det $\geq$ 1-class if the group G belongs to $\mathcal{G}$. But it is not correct that G satisfies the Approximation Conjecture for any normal chain $\{G_i\}$ if G belongs to $\mathcal{G}$; one additionally needs to know that each G/G_i belongs to $\mathcal{G}$.

Recall that the Determinant Conjecture 5.1 is known for a large class of groups; for instance, it is true for all sofic groups. Suppose that each quotient G/G_i is finite. Then we recover Theorem 6.2 from Remark 7.2.

A typical application of the Approximation Conjecture 7.1 is the following. If G satisfies the Approximation Conjecture 7.1 and for a given normal chain $\{G_i\}$ each G/G_i is torsion-free and satisfies the Atiyah Conjecture 3.1, then G is torsion-free and satisfies the Atiyah Conjecture 3.1, since a limit of a convergent sequence of integers is an integer again.

For more information about the Approximation Conjecture and its applications we refer to [**87**, Chapter 13] and [**116**].

7.2. Approximation of Fuglede–Kadison determinants. The following conjecture is taken from [**90**, Conjecture 14.1 on p. 308].

CONJECTURE 7.3 (Approximation Conjecture for Fuglede–Kadison determinants). *A group G satisfies the* Approximation Conjecture for Fuglede–Kadison determinants *if for any normal chain $\{G_i\}$ and any matrix $A \in M_{r,s}(\mathbb{Q}G)$ we get for the Fuglede–Kadison determinant*

$$\begin{aligned}\det_{\mathcal{N}(G)}\bigl(r_A^{(2)}\colon L^2(G)^r \to L^2(G)^s\bigr)\\ = \lim_{i\in I} \det_{\mathcal{N}(G/G_i)}\bigl(r_{A[i]}^{(2)}\colon L^2(G/G_i)^r \to L^2(G/G_i)^s\bigr).\end{aligned}$$

An equivalent chain complex version can be found in [**90**, Conjecture 15.3 on p. 309].

7.3. Approximation of L^2-torsion. Let $\overline{M}$ be a Riemannian manifold without boundary that comes with a proper free cocompact isometric G-action. Denote by $M[i]$ the Riemannian manifold obtained from $\overline{M}$ by dividing out the G_i-action. The Riemannian metric on $M[i]$ is induced by the one on M. There is an obvious proper free cocompact isometric G/G_i-action on $M[i]$ induced by the given G-action on $\overline{M}$. Notice that $M = \overline{M}/G$ is a closed Riemannian manifold and we get a G-covering $\overline{M} \to M$ and a G/G_i-covering $M[i] \to M$ which are compatible with the Riemannian metrics. Denote by

$$\rho_{\mathrm{an}}^{(2)}(\overline{M};\mathcal{N}(G)) \in \mathbb{R}; \tag{7.2}$$

$$\rho_{\mathrm{an}}^{(2)}(M[i];\mathcal{N}(G/G_i)) \in \mathbb{R}, \tag{7.3}$$

their *analytic L^2-torsion* over $\mathcal{N}(G)$ and $\mathcal{N}(G/G_i)$, respectively; see [**87**, Definition 3.128 on page 178]. Note that we do *not* require that $\overline{M}$ or $M[i]$ is L^2-acyclic. If $[G : G_i]$ is finite, $\rho_{\mathrm{an}}^{(2)}(M[i];\mathcal{N}(G/G_i))$ is the Ray–Singer torsion of the closed Riemannian manifold multiplied with $\frac{1}{[G:G_i]}$ defined in [**109**].

CONJECTURE 7.4 (Approximation Conjecture for analytic L^2-torsion). *A group G satisfies the* Approximation Conjecture for analytic L^2-torsion *if for any normal chain $\{G_i\}$ and Riemannian manifold $\overline{M}$ without boundary and with a proper free cocompact isometric G-action*

$$\rho_{\mathrm{an}}^{(2)}(\overline{M};\mathcal{N}(G)) = \lim_{i\in I} \rho_{\mathrm{an}}^{(2)}(M[i];\mathcal{N}(G/G_i)).$$

There are topological counterparts which we will denote by $\rho_{\mathrm{top}}^{(2)}(X[i];\mathcal{N}(G/G_i))$ and $\rho_{\mathrm{top}}^{(2)}(\overline{X};\mathcal{N}(G))$; see [**87**, Definition 3.120 on p. 176]. They agree with their analytic versions by [**21**]. So Conjecture 7.4 is equivalent to its topological counterpart.

CONJECTURE 7.5 (Approximation Conjecture for topological torsion). *A group G satisfies the* Approximation Conjecture for topological L^2-torsion *if, for any normal chain $\{G_i\}$ and Riemannian manifold $\overline{M}$ without boundary and with a proper free cocompact isometric G-action,*

$$\rho_{\mathrm{top}}^{(2)}(\overline{M};\mathcal{N}(G)) = \lim_{i\to\infty} \rho_{\mathrm{top}}(M[i];\mathcal{N}(G/G_I))$$

The next result is proved in [**90**, Theorem 15.6 on p. 310].

THEOREM 7.6. *Suppose that G satisfies the Approximation Conjecture* 7.3 *for Fuglede–Kadison determinants.*

Let $\overline{M}$ be a Riemannian manifold without boundary that comes with a proper free cocompact isometric G-action. Suppose that $b_n^{(2)}(\overline{M};\mathcal{N}(G)) = 0$ holds for all $n \geq 0$.

Then

$$\rho_{\text{an}}^{(2)}(\overline{M};\mathcal{N}(G)) = \lim_{i\in I} \rho_{\text{an}}^{(2)}(M[i];\mathcal{N}(G/G_i)).$$

REMARK 7.7 (Relating the Approximation Conjectures for Fuglede–Kadison determinant and torsion invariants). It is conceivable that Theorem 7.6 is still true, if we drop the assumption that $b_n^{(2)}(\overline{M};\mathcal{N}(G))$ vanishes for all $n \geq 0$, but our present proof works only under this assumption, see [**90**, Remark 16.2 on p. 315]. If we can drop this assumption in Theorem 7.6, then Theorem 7.6 just boils down to the statement that G satisfies equivalent Conjectures 7.4 and 7.5 about L^2-torsion, provided that G satisfies the Approximation Conjecture 7.3 for Fuglede–Kadison determinants.

REMARK 7.8 (Strategy of proof of Conjecture 7.3). A strategy to prove the Approximation Conjecture 7.3 for Fuglede–Kadison determinants is described in [**90**, Theorem 17.1 on page 316]. The key problem is isolated in the *uniform integrability condition* appearing in [**90**, Theorem 17.1 (v)]. Why it does not follow from known facts about spectral density functions is explained in [**90**, Lemma 17.2 on p. 318], whereas in [**90**, Remark 17.3 on p. 321] it is discussed why there is some hope that it may hold in many situations. In [**90**, Theorem 17.6 on p. 322] a sufficient condition for the validity of the uniform integrability condition is described.

REMARK 7.9 ((Very poor) status of Conjecture 7.3). Unfortunately, there is no infinite group G besides infinite virtually cyclic groups for which the Approximation Conjecture 7.3 for Fuglede–Kadison determinants is known to be true. It holds for finite groups G for trivial reasons. For $G = \mathbb{Z}$ (and hence for every infinite virtually cyclic group) Conjecture 7.3 has been proven by Schmidt [**118**]; see also [**87**, Lemma 13.53 on page 478] and [**91**, Lemma 7.25]. Note that the proof of this very special case already requires some input from the theory of diophantine equations, namely the estimate [**87**, 13.65] taken from [**123**, Corollary B1 on page 30].

We conclude from [**90**, Theorem 15.7] that in the situation of the Approximation Conjecture for Fuglede–Kadison determinants 7.3 we always have the inequality

$$\begin{aligned} (7.4)\quad & \dim_{\mathcal{N}(G)}\big(\ker\big(r_A^{(2)}\colon L^2(G)^r \to L^2(G)^s\big)\big) \\ & \geq \limsup_{i\to\infty} \dim_{\mathcal{N}(G/G_i)}\big(\ker\big(r_{A[i]}^{(2)}\colon L^2(G/G_i)^r \to L^2(G/G_i)^s\big)\big) \end{aligned}$$

provided that each quotient G/G_i satisfies the Determinant Conjecture 5.1. If $G = \mathbb{Z}^n$ and each G/G_i is finite, then the inequality (7.4) for the limit superior is known to be an equality by Lê [**71**, Theorem 3]; see also Raimbault [**107**].

REMARK 7.10. The relationship between Conjecture 6.11 about homological torsion growth and L^2-torsion and the Approximation Conjecture 7.5 for finite index normal chains is described in terms of regulators in [**90**, Sections 8–10].

7.4. More general setups. Some of the approximation conjectures above can also be formulated in more generality or a slightly different context. One can replace the normal chain $\{G_i\}$ indexed by $i \in \{0, 1, 2, \ldots\}$ by an inverse system $\{G_i \mid i \in I\}$ of normal subgroups of G directed by inclusion over the directed set I such that $\bigcap_{i \in I} G_i = \{1\}$; see [**90**, Section 13]. One may consider so-called *Farber sequences*, where the subgroups G_i are not necessarily normal but normal in an asymptotic sense; see [**2**, Section 10]. There is the notion of Benjamini–Schramm convergence, see [**1**, Definition 1.1]. Or one can consider convergence in the space or marked groups; see [**58**, Section 1.3].

8. L^2-invariants and simplicial volume

The simplicial volume of a manifold is a topological variant of the (Riemannian) volume which agrees with it for hyperbolic manifolds up to a dimension constant and was introduced by Gromov [**44**].

CONJECTURE 8.1 (Simplicial volume and L^2-invariants). *Let M be an aspherical closed orientable manifold of dimension ≥ 1. Suppose that its simplicial volume $\|M\|$ vanishes. Then $\widetilde{M}$ is of determinant class and*

$$b_p^{(2)}(\widetilde{M}) = 0 \quad \textit{for} \quad p \geq 0;$$
$$\rho^{(2)}(\widetilde{M}) = 0.$$

Gromov first asked in [**48**, Section 8A on p. 232] whether under the conditions in Conjecture 8.1 the Euler characteristic of M vanishes, and notes that in all available examples even the L^2-Betti numbers of M vanish. The part about L^2-torsion appears in [**84**, Conjecture 3.2].

Conjecture 8.1 is discussed in detail in [**87**, Chapter 14]. No essential breakthrough has been made since then, but there are some interesting papers connected to this problem, such as [**37, 112, 113**]. So far, all evidence for Conjecture 8.1 has been computational, but there is no structural reason known for it. It is intriguing, since it relates rather different invariants to one another and is one of the typical conjectures, which only make sense for closed manifolds when they are aspherical.

For an aspherical closed orientable manifold of dimension ≥ 1 its simplicial volume $\|M\|$ vanishes if $\pi_1(M)$ is amenable, see [**44**, p. 40] and [**54**, Theorem 4.3]. However, it is not known whether $\|M\|$ vanishes if $\pi_1(M)$ contains a normal infinite amenable subgroup or at least an elementary amenable infinite normal subgroup. Recall that $\rho^{(2)}(\widetilde{M})$ vanishes if $\pi_1(M)$ contains a normal infinite elementary amenable subgroup (see (11) in Section 2.5) but it is not known wether $\rho^{(2)}(\widetilde{M})$ vanishes if $\pi_1(M)$ contains a normal infinite amenable subgroup.

As an approach to Conjecture 8.1, Gromov suggested in [**49**] to use a new invariant, the integral foliated simplicial volume, to estimate L^2-Betti numbers from above. This was carried out in [**119**, Corollary 5.28]. However, it is still open whether vanishing of simplicial volume implies vanishing of integral foliated simplicial volume (of oriented closed aspherical manifolds).

9. L^2-invariants of groups

Recall that L^2-Betti numbers $b_n^{(2)}(G)$ of a group G were defined in (2.6). We call a group G *admissible*, if there exists a finite CW-model BG for its classifying space, we have $b_n^{(2)}(EG;\mathcal{N}(G)) = 0$ for $n \geq 0$, and G satisfies the Determinant Conjecture 5.1. We define the *L^2-torsion of an admissible group G*

$$\rho^{(2)}(G) = \rho^{(2)}(EG;\mathcal{N}(G)) \in \mathbb{R} \tag{9.1}$$

by the L^2-torsion of $EG = \widetilde{BG}$. Since two models for EG are G-homotopy equivalent and G satisfies the Determinant Conjecture 5.1, the real number $\rho^{(2)}(EG;\mathcal{N}(G))$ is up to a constant well defined and is independent of the choice of the model for EG. Hence the notion of the L^2-torsion of an admissible group in (9.1) makes sense.

9.1. Vanishing results. Some vanishing criterions for the L^2-Betti numbers $b_n^{(2)}(G)$ of a group can be found in [**87**, Theorem 7.2 on p. 294 and Theorem 7.4 on p. 295]. For instance, if G contains a normal subgroup H such that $b_n^{(2)}(H) = 0$ for all $n \leq d$ for some fixed natural number d, then $b_n^{(2)}(G) = 0$ for all $n \leq d$. If G contains an amenable infinite normal subgroup, then $b_n^{(2)}(G) = 0$ holds for all $n \geq 0$. If G occurs as an extension $1 \to H \to G \to Q \to 1$ of infinite groups such that H is finitely generated or, more generally, satisfies $b_1^{(2)}(H) < \infty$, then $b_1^{(2)}(G) = 0$. See [**111**] and [**114**] for generalizations of the last assertion. If G is admissible and contains an elementary amenable infinite normal subgroup, then $\rho^{(2)}(G) = 0$ holds; see (11) in Section 2.5. A very interesting interaction between the notions of L^2-torsion and entropy is developed in [**72**] and used to show that for any admissible group which is amenable, $\rho^{(2)}(G) = 0$ holds; see [**72**, Theorem 1.3]. Gaboriau and Seward [**42**, Corollaries 1.2 and 5.9] relate L^2-Betti numbers of groups to the failure of the Yuzvinsky addition formula for nonamenable groups. It is unknown whether $\rho^{(2)}(G) = 0$ vanishes if G contains an amenable infinite normal subgroup.

REMARK 9.1. A combinatorial computation of L^2-invariants is described in [**87**, Section 3.7]. Given a matrix $A \in M_{m,n}(\mathbb{C}G)$, one can assign to it its *characteristic sequence* $\{c(A,K)_n\}$ for some large enough number K. The sequence $\{c(A,K)_n\}$ is monotone decreasing sequence of nonnegative real numbers that converges to the von Neumann dimension $\dim_{\mathcal{N}(G)}(\ker(R_A))$ of the kernel of the induced $\mathcal{N}(G)$-homomorphism $R_A\colon \mathcal{N}(G)^m \to \mathcal{N}(G)^n$. There is some control over the speed of convergence, which is often very fast, actually exponentially in n. Analogously, there is a way of computing the Fuglede–Kadison determinant of R_A in terms of the characteristic sequence. Given an algorithm to decide the word problem in G, one obtains an algorithm to compute the characteristic sequence. Unfortunately this algorithm seems to be exponentially running. But at least each $c(A,K)_n$ gives an upper bound for $\dim_{\mathcal{N}(G)}(R_A)$. This may be useful in view of the Atiyah Conjecture 3.1 to prove the vanishing of L^2-Betti numbers $b_n^{(2)}(\widetilde{X})$ for a finite CW-complex X for which there is an upper bound on the orders of the finite subgroups of $\pi_1(X)$. See also Remark 3.7.

9.2. The first L^2-Betti number and applications to group theory. The vanishing of $b_1^{(2)}(G)$ has some interesting consequences, provided that G is finitely presented. Namely, it implies that the deficiency of G is bounded from above by 1 and that for any oriented closed manifold M of dimension four the inequality $|\operatorname{sign}(M)| \leq \chi(M)$ holds for its signature $\operatorname{sign}(M)$ and its Euler characteristic $\chi(M)$; see [**83**, Theorems 5.1 and 6.1]. The following result is due to Kielak [**63**, Theorem 5.3] and generalizes the work of Agol [**5**]

THEOREM 9.2. *Let G be an infinite finitely generated group which is virtually* RFRS, *where* RFRS *stands for residually finite rationally solvable.*

Then G is virtually fibered, in the sense that it admits a finite index subgroup mapping onto $\mathbb{Z}$ with a finitely generated kernel, if and only if $b_1^{(2)}(G) = 0$ holds.

There are relations between the nonvanishing of $b_1^{(2)}(G)$ and the question whether the finitely presented group G is *large*, i.e, has a subgroup of finite index which maps surjectively to a non-Abelian free group; see, for instance, [**69**, Theorem 1.4], [**70**, Theorem 1.6].

The following questions are taken from [**90**, Section 4], where more explanations and references to the literature, such as [**3, 4, 20, 31, 39–41, 68, 94, 104, 117**], are also given.

QUESTION 9.3 (Rank gradient, cost, and first L^2-Betti number). *Let G be an infinite finitely generated residually finite group. Let $(G_i)_{i \geq 0}$ be a descending chain of normal subgroups of finite index of G with $\bigcap_{i \geq 0} G_i = \{1\}$.*

Do we have

$$b_1^{(2)}(G) = \operatorname{cost}(G) - 1 = \operatorname{RG}(G; (G_i)_{i \geq 0})?$$

QUESTION 9.4 (Rank gradient, cost, first L^2-Betti number, and approximation). *Let G be a finitely presented infinite residually finite group. Let (G_i) be a descending chain of normal subgroups of finite index of G with $\bigcap_{i \geq 0} G_i = \{1\}$. Let F be any field.*

Do we have

$$\lim_{i \to \infty} \frac{b_1(G_i; F) - 1}{[G : G_i]} = b_1^{(2)}(G) - b_0^{(2)}(G) = \operatorname{cost}(G) - 1 = \operatorname{RG}(G; (G_i)_{i \geq 0})?$$

Note that a positive answer to the questions above also includes the statement, that $\lim_{i \to \infty} \frac{b_1(X[i]; F)}{[G:G_i]}$ and $\operatorname{RG}(G; (G_i)_{i \geq 0})$ are independent of the chain and the characteristic of F. It is possible that the answer is positive also in the case, where the characteristic of F is not zero, since in Remark 6.6, the counterexamples of Avramidi, Okun, and Schreve [**13**, Theorem 1] do not work in degree $n = 1$.

QUESTION 9.5. *Let G be a finitely generated group. Do we have*

$$b_1^{(2)}(G) = \operatorname{cost}(G) - 1$$

and has the the Fixed Price Problem a positive answer? (The Fixed Price Problem is to decide whether the cost of every standard action of G, i.e., an essentially free G-action on a standard Borel space with G-invariant probability measure, is equal to the cost of G; see [**39**, Question 1.8]*.)*

Higher rank versions of the rank gradient are discussed in [**90**, Section 5].

9.3. The L^2-torsion and applications to group theory. As we have explained above, L^2-Betti numbers have been exploited for group theory. We think that there is a lot of potential for the L^2-torsion of a group G to have striking applications to group theory, and we encourage group theorists to work on these types of questions.

A typical question is the following. If M is a closed hyperbolic manifold of odd dimension, then $\rho^{(2)}(\pi_1(M))$ is, up to a constant depending only on the dimension, equal to the volume of M, see [**52, 97**]. Since $\pi_1(M)$ is a word-hyperbolic group, one may ask what $\rho^{(2)}(G)$ measures for a det-L^2-acyclic torsionfree word-hyperbolic group.

We want to mention the following invariant of an automorphism $f\colon BG \to BG$ of a group G for which there exists a finite model for BG and which satisfies the Determinant Conjecture 5.1. Then $G \rtimes_f \mathbb{Z}$ is admissible, and hence we can define the L^2-torsion of f by

$$\rho^{(2)}(f) := \rho^{(2)}(G \rtimes_f \mathbb{Z}). \tag{9.2}$$

This invariant has the following properties; see [**87**, Theorem 7.27 on page 305].

THEOREM 9.6. *Suppose that all groups appearing below have finite CW-models for their classifying spaces and satisfy the Determinant Conjecture* 5.1.

(1) *Suppose that G is the amalgamated product $G_1 *_{G_0} G_2$ for subgroups $G_i \subset G$ and the automorphism $f\colon G \to G$ is the amalgamated product $f_1 *_{f_0} f_2$ for automorphisms $f_i\colon G_i \to G_i$. Then*
$$\rho^{(2)}(f) = \rho^{(2)}(f_1) + \rho^{(2)}(f_2) - \rho^{(2)}(f_0);$$
(2) *Let $f\colon G \to H$ and $g\colon H \to G$ be isomorphisms of groups. Then*
$$\rho^{(2)}(f \circ g) = \rho^{(2)}(g \circ f).$$
In particular, $\rho^{(2)}(f)$ is invariant under conjugation with automorphisms;
(3) *Suppose that the following diagram of groups*
$$\begin{array}{ccccccccc} 1 & \longrightarrow & G_1 & \xrightarrow{i} & G_2 & \xrightarrow{p} & G_3 & \longrightarrow & 1 \\ & & \downarrow{\scriptstyle f_1} & & \downarrow{\scriptstyle f_2} & & \downarrow{\scriptstyle \mathrm{id}} & & \\ 1 & \longrightarrow & G_1 & \xrightarrow{i} & G_2 & \xrightarrow{p} & G_3 & \longrightarrow & 1 \end{array}$$
commutes, has exact rows and its vertical arrows are automorphisms. Then
$$\rho^{(2)}(f_2) = \chi(BG_3) \cdot \rho^{(2)}(f_1);$$
(4) *Let $f\colon G \to G$ be an automorphism of a group. Then for all integers $n \geq 1$*
$$\rho^{(2)}(f^n) = n \cdot \rho^{(2)}(f);$$
(5) *We have $\rho^{(2)}(f) = 0$, if G satisfies one of the following conditions:*
   (a) *We have $b_n^{(2)}(G) = 0$ for every $n \geq 0$;*
   (b) *G contains an amenable infinite normal subgroup.*

Let S be a compact connected orientable 2-dimensional manifold, possibly with boundary. Let $f\colon S \to S$ be an orientation preserving homeomorphism. The mapping torus T_f is a compact connected orientable 3-manifold, whose boundary is empty or a disjoint union of 2-dimensional tori. Then there is a maximal family of embedded incompressible tori, which are pairwise not isotopic and not boundary parallel, such that it decomposes T_f into pieces that are Seifert or hyperbolic. Let M_1, M_2, ..., M_r be the hyperbolic pieces. They all have finite volume $\operatorname{vol}(M_i)$. The following result is taken from [**87**, Theorem 7.28 on page 307]

THEOREM 9.7. *If S is S^2, D^2, or T^2, then $\rho^{(2)}(f) = 0$. Otherwise we get*
$$\rho^{(2)}(\pi_1(f)\colon \pi_1(S) \to \pi_1(S)) = \frac{-1}{6\pi} \cdot \sum_{i=1}^{r} \operatorname{vol}(M_i).$$

The combinatorial approach for the computation of $\rho^{(2)}(f)$ in terms of characteristic sequences of Remark 9.1 is described in detail in [**87**, Subsection 7.34]. The favorite and so far unexplored case is when G is a finitely generated free group.

QUESTION 9.8. *Let G be finitely generated non-Abelian free group.*

(1) *Is $\rho^{(2)}(f) \leq 0$ for any automorphism f of G;*
(2) *What is the structure of the countable set $\{\rho^{(2)}(f)\}$, where f runs through the automorphisms of G?*
(3) *Given a real number $r < 0$, is the set of conjugacy classes of fully irreducible automorphisms f of G with $\rho^{(2)}(f) = r$ finite?*

We will discuss measure equivalence in Section 10 and twisted L^2-torsion in Section 12.

10. Measure equivalence

Gaboriau [**40**] introduced *L^2-Betti numbers of measured equivalence relations* and proved that two measure equivalent countable groups have proportional L^2-Betti numbers. This notion turned out to have many important applications to the theory of von Neumann algebras in recent years, most notably through the work of Popa [**106**].

The notion of *measure equivalence* was introduced by Gromov [**48**, 0.5.E].

DEFINITION 10.1. Two countable groups G and H are called *measure equivalent with index* $c = I(G,H) > 0$ if there exists a nontrivial standard measure space (Ω, μ) on which $G \times H$ acts such that the restricted actions of $G = G \times \{1\}$ and $H = \{1\} \times H$ have measurable fundamental domains $X \subset \Omega$ and $Y \subset \Omega$, with $\mu(X) < \infty$, $\mu(Y) < \infty$, and $c = \mu(X)/\mu(Y)$. The space (Ω, μ) is called a *measure coupling* between G and H (of index c).

The following conjecture is taken from [**96**, Conjecture 1.2].

CONJECTURE 10.2. *Let G and H be two admissible groups, which are measure equivalent with index $I(G,H) > 0$. Then*

$$\rho^{(2)}(G) = I(G,H) \cdot \rho^{(2)}(H).$$

Due to Gaboriau [**40**], the vanishing of the nth L^2-Betti number $b_n^{(2)}(G)$ is an invariant of the measure equivalence class of a countable group G. If all L^2-Betti numbers vanish and G is an admissible group, then the vanishing of the L^2-torsion is a secondary invariant of the measure equivalence class of a countable group G provided that Conjecture 10.2 holds.

Evidence for Conjecture 10.2 comes from [**96**, Conjecture 1.10] which says that Conjecture 10.2 is true if we replace measure equivalence by the stronger notion of uniform measure equivalence, see [**96**, Definition 1.3], and assume that G satisfies the Measure Theoretic Determinant Conjecture; see [**96**, Conjecture 1.7].

11. Zero-in-the-Spectrum Conjecture

The next conjecture appears for the first time in Gromov's article [**45**, page 120].

CONJECTURE 11.1 (Zero-in-the-Spectrum Conjecture). *Suppose that $\widetilde{M}$ is the universal covering of the aspherical closed Riemannian manifold M (with the Riemannian metric coming from M). Then for some $p \geq 0$ zero is in the spectrum of the minimal closure*

$$(\Delta_p)_{\min}\colon \operatorname{dom}((\Delta_p)_{\min}) \subset L^2\Omega^p(\widetilde{M}) \to L^2\Omega^p(\widetilde{M})$$

of the Laplacian acting on smooth p-forms on $\widetilde{M}$.

The Zero-in-the-Spectrum Conjecture 11.1 is known to be true if one of the following conditions is satisfied:

(1) $\dim(M) \leq 3$;
(2) M is a locally symmetric space;
(3) M possesses a Riemannian metric whose sectional curvature is nonpositive;
(4) M is an aspherical closed Kähler manifold whose fundamental group is word-hyperbolic in the sense of [**46**];
(5) $\pi_1(M)$ satisfies the strong Novikov Conjecture.

If one drops the conditions "aspherical" in the Zero-in-the-Spectrum Conjecture 11.1, then there are counterexamples.

For the proofs of the claim above and the relevant reference, such as [**32**, **45**, **81**], and for more information about the Zero-in-the-Spectrum Conjecture 11.1, we refer to [**87**, Chapter 12].

12. Twisting with finite-dimensional representations

A prominent open problem is whether one can twist L^2-invariants with (not necessarily unitary) finite-dimensional representations. A basic and systematical treatment of this problem can be found in [**91**].

For L^2-Betti numbers there is the conjecture that this just boils down to multiplying the untwisted L^2-Betti number with the dimension of the representation; see [**19**, Conjecture 2] and [**91**, Question 0.1]. This conjecture is proved for sofic groups by Boschheidgen, Jaikin, and Zapirian[**19**, Theorem 1.1] and for locally indicable groups by Kielak and Sun [**65**, Theorem 4.5].

For L^2-torsion the effect of the twisting is much more interesting and this leads to new invariants. Especially in dimension 3 there has been made a lot of progress, and interesting open problems occur. For instance, there are interesting relations between the Alexander-torsion function, which is given by twisting the L^2-torsion with a family of one-dimensional representations associated to an element in $H^1(M;\mathbb{Z})$, and the Thurston polytope. A prominent open problem is whether the regulare Fuglede–Kadison determinant is continuous; see [**91**, Question 9.11]. For more information and the relevant references, such as [**28**, **34**–**36**, **38**, **62**, **73**, **78**], we refer to [**91**, Section 10] and the survey article [**92**].

13. Group von Neumann algebras over $\mathbb{F}_p$

Some of the problems discussed above are of the shape that we understand the case of a field of characteristic zero well using the group von Neumann algebra $\mathcal{N}(G)$, but do not know what happens in prime characteristic p. A prominent example are the questions about approximation of L^2-Betti numbers in prime characteristic; see Section 6.5. It seems to be conceivable that the relevant limits exists and are independent of the chains, but the value of the limits depend on whether we work in characteristic zero or in prime characteristic. So the original hope that one always gets as a limit the L^2-invariants defined in terms of the von Neumann algebra also in prime characteristic turns out not to be fulfilled. So we face the new problem what the limit could be in the prime characteristic case.

This raises the question whether there is an $\mathbb{F}_p$-analog of the group von Neumann algebra. In order to treat L^2-Betti numbers in prime characteristic, one would hope for the existence of a $\mathbb{F}_p$-algebra $\mathcal{N}(G;p)$ which contains the group ring $\mathbb{F}_pG$ and comes with a dimension function for finitely generated projective $\mathcal{N}(G;p)$-modules satisfying [**87**, Assumption 6.2 on p. 238]. Then [**87**, Definition 6.6 and Theorem 6.7 on p. 239] would apply, and one would get a dimension function for all $\mathcal{N}(G;p)$-modules which has many useful features. Finally, one would define the nth L^2-Betti number in characteristic p of a G-space X to be

$$b_n^{(2)}(Y;\mathcal{N}(G;p)) := \dim_{\mathcal{N}(G;p)}\big(H_n(\mathcal{N}(G;p)\otimes_{\mathbb{F}_pG}C_*^s(Y;\mathbb{F}_p))\big) \in \mathbb{R}^{\geq 0}\amalg\{\infty\}, \tag{13.1}$$

where $C_*^s(Y;\mathbb{F}_p)$ is the $\mathbb{F}_pG$-chain complex given by the singular chain complex with coefficient in $\mathbb{F}_p$. For a group G one would define its nth L^2-Betti number in characteristic p

$$b_n^{(2)}(G;p) = b_n^{(2)}(EG;\mathcal{N}(G;p)). \tag{13.2}$$

The hope is that then the corresponding sequences of normalized Betti numbers with coefficients in $\mathbb{F}_p$ converge for any normal chain to the L^2-Betti numbers with coefficients in $\mathbb{F}_p$; see also Remark 6.7.

14. L^2-invariants and condensed mathematics

Condensed mathematics, a theory recently developed by Clausen and Scholze [**120–122**], is a framework which aims to remedy ill behavior of the category of topological spaces with respect to algebraic structure, for example, the fact that the category of topological Abelian groups is not an Abelian category. It has proven useful in incorporating geometric and analytic structures that appear in arithmetic, or complex geometry and we are far from understanding its full capabilities. In the context of L^2-invariants, condensed mathematics poses at least two interesting questions.

(1) The theory of L^2-invariants heavily relies on techniques mixing algebra and topology or functional analysis. Can condensed mathematics help to extend the current formalism of L^2-invariants in a way which sheds light on open problems in the area? For example, are there analogs of the group von Neumann algebra of a discrete group over $\mathbb{F}_p$?
(2) Can L^2-invariants be applied to a wider class of geometric problems coming from condensed mathematics?

In this section we attempt to make a first step towards the second question by defining L^2-Betti numbers of condensed sets carrying an action of a discrete group in three different ways that extend the current definition for nicely behaved spaces, e.g., CW-complexes:

(1) homology through solidification;
(2) condensed cohomology;
(3) condensed singular homology.

We would hope that more sophisticated setups to associate L^2-Betti numbers to objects in analytic geometry lead to new interesting invariants. The reader should be aware that we are freely using the language of ∞-categories.

Let us begin by giving a brief summary of the main definitions from condensed mathematics necessary for this purpose. For more details and proofs, we refer the reader to Peter Scholze's lecture notes [**120**–**122**]. We also ignore set theoretic size issues, which are treated with more care in aforementioned references.

Denote by edCH the category of extremally disconnected compact Hausdorff spaces (extremally disconnected, meaning that the closure of any of its open subsets is again open) and continuous maps. It can be made into a site with coverings given by finite families of jointly surjective maps. For C, a category with finite limits, the category $\mathsf{Cond}(\mathsf{C})$ of condensed objects in C is defined as the category of sheaves on edCH with values in C. It is not hard to check that this identifies with the full subcategory of $\mathrm{Fun}(\mathsf{edCH}^{\mathrm{op}}, \mathsf{C})$ of finite product preserving functors. Explicitly, a condensed object in C is a functor

$$X\colon \mathsf{edCH}^{\mathrm{op}} \to \mathsf{C}$$

which satisfies

(1) $X(\emptyset)$ is the terminal object in C;
(2) For any two objects T_1 and T_2 in edCH the natural map
$$X(T_1 \amalg T_2) \to X(T_1) \times X(T_2)$$
is an isomorphism.

The category $\mathsf{Cond}(\mathsf{Set})$ of condensed sets is an enlargement of the category of compactly generated topological spaces in the following sense.

Proposition 14.1 ([**121**, Proposition 1.7]). *The functor* $\mathsf{Top} \to \mathsf{Cond}(\mathsf{Set})$, $X \mapsto \underline{X}$ *given by the restricted Yoneda embedding admits a left adjoint and is fully faithful when restricted to compactly generated topological spaces.*

The nicely behaved enlargement of topological Abelian groups is now given by the category $\mathsf{Cond}(\mathsf{Ab})$ of condensed Abelian groups. It is a complete and cocomplete Abelian category additionally satisfying Grothendieck's axioms (AB5) and (AB6). We now summarize some additional structure:

(1) The forgetful functor $\mathsf{Cond}(\mathsf{Ab}) \to \mathsf{Cond}(\mathsf{Set})$ admits a left adjoint $X \mapsto \mathbb{Z}[X]$.
(2) The category $\mathsf{Cond}(\mathsf{Ab})$ admits a closed symmetric monoidal structure with $M \otimes N$ given by the sheafification of the presheaf $T \mapsto M(T) \otimes_{\mathbb{Z}} N(T)$.
(3) There is an equivalence $D(\mathsf{Cond}(\mathsf{Ab})) \simeq \mathsf{Cond}(D(\mathbb{Z}))$ between the derived ∞-category of condensed abelian groups and condensed objects in the derived ∞-category of $\mathbb{Z}$-modules. The category $\mathsf{Cond}(D(\mathbb{Z}))$ is a stable presentably symmetric monoidal ∞-category with symmetric monoidal structure induced from the one on $D(\mathbb{Z})$. It can also be identified with the left derived tensor product on condensed abelian groups.

We now want to outline approach (1) homology through solidification to define L^2-Betti numbers of condensed sets. First, we recall some background on solid Abelian groups from [**121**, Lecture 5 and 6].

Definition 14.2. For a profinite set T define

$$\mathbb{Z}[T]^{\blacksquare} = \underline{\mathrm{Hom}}_{\mathsf{Cond}(\mathsf{Ab})}(\underline{C(T, \mathbb{Z})}, \mathbb{Z})$$

where $C(T,\mathbb{Z})$ denotes the space of continuous maps endowed with the compact open topology. It comes with a canonical map $\mathbb{Z}[T] \to \mathbb{Z}[T]^{\blacksquare}$. Equivalently, if $T = \lim_i T_i$ is given as the cofiltered limit of finite sets, one can identify $\mathbb{Z}[T]^{\blacksquare} \simeq \lim_i \mathbb{Z}[T_i]$ where the limit is formed in $\mathsf{Cond}(\mathsf{Ab})$.

A condensed Abelian group M is then called *solid* if for all profinite sets T the induced map

$$\mathrm{Hom}_{\mathsf{Cond}(\mathsf{Ab})}(\mathbb{Z}[T]^{\blacksquare}, M) \to \mathrm{Hom}_{\mathsf{Cond}(\mathsf{Ab})}(\mathbb{Z}[T], M)$$

is an equivalence. Denote by $\mathsf{Solid}(\mathsf{Ab}) \subset \mathsf{Cond}(\mathsf{Ab})$ the full subcategory of solid abelian groups.

Similarly, a derived condensed Abelian group $C \in D(\mathsf{Cond}(\mathsf{Ab})) \simeq \mathsf{Cond}(D(\mathbb{Z}))$ is called *solid* if for all profinite sets T the induced map

$$\mathrm{Hom}_{D(\mathsf{Cond}(\mathsf{Ab}))}(\mathbb{Z}[T]^{\blacksquare}, C) \to \mathrm{Hom}_{D(\mathsf{Cond}(\mathsf{Ab}))}(\mathbb{Z}[T], C)$$

is an equivalence in $D(\mathbb{Z})$. Denote by $\mathsf{Solid}(D(\mathbb{Z})) \subseteq \mathsf{Cond}(D(\mathbb{Z}))$ the full subcategory of solid derived $\mathbb{Z}$-modules.

The following theorem is the main result from [**121**, Lecture 5].

THEOREM 14.3. *The full inclusion* $\mathsf{Solid}(D(\mathbb{Z})) \subseteq \mathsf{Cond}(D(\mathbb{Z}))$ *admits a left adjoint* $(-)^{L\blacksquare}$ *called the solidification. Furthermore, a condensed derived* $\mathbb{Z}$*-module is solid if and only if all of its homology groups are solid Abelian groups.*

Now we come to the definition of L^2-Betti numbers of condensed sets. Let X be a condensed set with G-action, i.e., an element in $\mathsf{Cond}(\mathsf{Set}^{BG}) \simeq \mathsf{Cond}(\mathsf{Set})^{BG}$. Consider the following composition

$$\begin{aligned}(14.1)\qquad & \mathsf{Cond}(\mathsf{Set})^{BG} \xrightarrow{\mathbb{Z}[-]} \mathsf{Cond}(D(\mathbb{Z}))^{BG} \xrightarrow{(-)^{L\blacksquare}} \mathsf{Cond}(D(\mathbb{Z}))^{BG} \\ & \simeq \mathsf{Cond}(D(\mathbb{Z})^{BG}) \simeq \mathsf{Cond}(D(\mathbb{Z}[G]))\end{aligned}$$

and denote the image of X under (14.1) by $\mathbb{Z}[X]^{L\blacksquare}$.

DEFINITION 14.4. Let R be a $\mathbb{Z}[G]$-algebra and X a condensed set with G-action. We define solid equivariant homology of X with coefficients in R by

$$H_n^{G,\blacksquare}(X;R) := \Gamma(H_n(\mathbb{Z}[X]^{L\blacksquare} \otimes_{\mathbb{Z}[G]} R)). \tag{14.2}$$

Here, $\Gamma\colon \mathsf{Cond}(\mathsf{Mod}_R) \to \mathsf{Mod}_R$ is the global sections functor and $- \otimes_{\mathbb{Z}[G]} R\colon D(\mathbb{Z}[G]) \to D(R)$ denotes the (derived) base change.

Furthermore, we define the L^2-Betti numbers of X by

$$b_n^{(2)}(X;\mathcal{N}(G)) := \dim_{\mathcal{N}(G)}(H_n^{G,\blacksquare}(X;\mathcal{N}(G))). \tag{14.3}$$

For free G-CW-complexes, we now compare equivariant homology and L^2-Betti numbers from Definition 14.4 with the classical ones, keeping Proposition 14.1 in mind. For this we need one further ingredient, giving a description of the singular chain complex of a CW-complex in the condensed world.

PROPOSITION 14.5 ([**121**, Example 6.5]). *There is an equivalence of the two functors*

$$\mathsf{CW} \xrightarrow{(-)} \mathsf{Cond}(\mathsf{Set}) \xrightarrow{\mathbb{Z}[-]} \mathsf{Cond}(D(\mathbb{Z})) \xrightarrow{(-)^{L\blacksquare}} \mathsf{Cond}(D(\mathbb{Z})) \text{ and}$$

$$\mathsf{CW} \xrightarrow{C_\bullet(-)} D(\mathbb{Z}) \xrightarrow{(-)} \mathsf{Cond}(D(\mathbb{Z})),$$

where $\mathsf{CW} \subseteq \mathsf{Top}$ *denotes the full subcategory of CW-complexes,* $C_\bullet(X)$ *denotes the singular chain complex of* X*, and the functor* $\underline{(-)}\colon D(\mathbb{Z}) \to \mathsf{Cond}(D(\mathbb{Z}))$ *sends an object* M *to the sheafification of the constant presheaf with value* M*.*

We now explain why this generalizes the L^2-Betti number of G-CW-complexes defined in the body of this article.

THEOREM 14.6. *Let* X *be a* G*-space which is* G*-homotopy equivalent to a* G*-CW-complex and* R *a* $\mathbb{Z}[G]$*-algebra. Then Borel homology of* X *with coefficients in* R *and homology of* X *in the sense of Definition* 14.4 *are equivalent. In particular, if* X *is* G*-homotopy equivalent to a free* G*-CW-complex, the* L^2*-Betti numbers of* X *in the sense of* (2.3) *and Definition* (14.3) *agree.*

PROOF. Consider the diagram
(14.4)

$$\begin{array}{ccccccc} \mathsf{CW}^{BG} & \xrightarrow{C_\bullet(-)} & D(\mathbb{Z})^{BG} & \xrightarrow{\simeq} & D(\mathbb{Z}[G]) & \xrightarrow{-\otimes_{\mathbb{Z}[G]} R} & D(R) \\ \downarrow{\scriptstyle \mathbb{Z}[-]} & & \downarrow{\scriptstyle \underline{(-)}} & & \downarrow{\scriptstyle \underline{(-)}} & & \downarrow{\scriptstyle \underline{(-)}} \\ \mathsf{Cond}(D(\mathbb{Z}))^{BG} & \xrightarrow{\underline{(-)}^{L\blacksquare}} & \mathsf{Cond}(D(\mathbb{Z}))^{BG} & \xrightarrow{\simeq} & \mathsf{Cond}(D(\mathbb{Z}[G])) & \xrightarrow{-\otimes_{\mathbb{Z}[G]} R} & \mathsf{Cond}(D(R)). \end{array}$$

The left square commutes by Proposition 14.5 and the rest of the diagram commutes for obvious reasons. The composite of the upper horizontal arrows recovers Borel homology of a G-CW-complex. It remains to note that taking global sections $\Gamma\colon \mathsf{Cond}(D(R)) \to D(R)$ is left inverse to the functor $\underline{(-)}\colon D(R) \to \mathsf{Cond}(D(R))$.

For a free G-CW-complex X, Borel homology of X with coefficients in R can be computed as homology of the degreewise tensor product of the singular chain complex of X with R, as used in the definition of L^2-Betti numbers in [**87**, Definition 6.50 on p. 263].

To extend this result to all G-spaces G-homotopy equivalent to free G-CW-complexes, observe that the functor $\mathsf{Top} \to \mathsf{Cond}(D(\mathbb{Z})), X \mapsto \mathbb{Z}[\underline{X}]^{L\blacksquare}$ sends homotopic maps to homotopic maps which formally follows from $\mathbb{Z}[[0,1]]^{L\blacksquare} \simeq \mathbb{Z}$. □

REMARK 14.7 (General actions). Note that the above definition does not recover L^2-Betti numbers for nonfree G-CW-complexes. To remedy this, one can work parametrised over the orbit category $\mathsf{Or}(G)$ instead of BG.

We now briefly explain the two alternative definitions of L^2-Betti numbers of condensed sets with G-action mentioned in the beginning of Section 14.

REMARK 14.8 (Condensed singular homology). One alternative is to model the construction of singular homology inside condensed sets defined as homology of the simplicial set

$$\mathrm{Hom}_{\mathsf{Cond(Set)}}(\underline{\Delta^\bullet}, X).$$

This in fact agrees with singular homology for all topological spaces since the spaces Δ^n are compactly generated so that the realization of the condensed set $\underline{\Delta^n}$ is Δ^n.

REMARK 14.9 (A cohomological variant). One can also define equivariant cohomology of condensed G-sets analogous to Definition 14.4. For this, one assigns to a condensed set X with G-action the homology of

$$\mathrm{Hom}_{\mathsf{Cond}(D(\mathbb{Z}[G]))}(\mathbb{Z}[X]^{L\blacksquare}, \underline{R}) \in D(R).$$

Then the same arguments as in Theorem 14.6 show that for free G-CW-complexes this recovers Borel cohomology.

Note that as for any Abelian group the induced constant condensed Abelian group is solid we have

$$\begin{aligned}\mathrm{Hom}_{\mathsf{Cond}(D(\mathbb{Z}[G]))}(\mathbb{Z}[X]^{L\blacksquare}, \underline{R}) &\simeq \mathrm{Hom}_{\mathsf{Cond}(D(\mathbb{Z}))}(\mathbb{Z}[X]^{L\blacksquare}, \underline{R})^{hG}\\ &\simeq \mathrm{Hom}_{\mathsf{Cond}(D(\mathbb{Z}))}(\mathbb{Z}[X], \underline{R})^{hG}\\ &\simeq \mathrm{Hom}_{\mathsf{Cond}(D(\mathbb{Z}[G]))}(\mathbb{Z}[X], \underline{R}).\end{aligned}$$

Because of that, this version of equivariant condensed cohomology looks most natural to us. From the viewpoint of L^2-invariants, working with homology instead of cohomology is often easier (for instance homology behaves better with respect to colimits) and chose to present the dual version in more detail.

Acknowledgments

This paper is funded by the ERC Advanced Grant "KL2MG-interactions" (no. 662400) of the third author granted by the European Research Council and by the Deutsche Forschungsgemeinschaft under Germany's Excellence Strategy - GZ 2047/1, Projekt-ID 390685813, Hausdorff Center for Mathematics at Bonn. All three authors are supported by the Max Planck Institute for Mathematics in Bonn.

We thank Grigori Avramidi for comments on an earlier version of this article and Emma Brink for helpful discussions concerning Section 14. Moreover, we thank the two referees for their valuable reports.

References

[1] M. Abert, N. Bergeron, I. Biringer, T. Gelander, N. Nikolov, J. Raimbault, and I. Samet, *On the growth of L^2-invariants for sequences of lattices in Lie groups*, Ann. of Math. (2) **185** (2017), no. 3, 711–790, DOI 10.4007/annals.2017.185.3.1. MR3664810

[2] M. Abert, N. Bergeron, M. Fraczyk, and D. Gaboriau, *On homology torsion growth*, Preprint, `arXiv:2106.13051`, 2021.

[3] M. Abért, A. Jaikin-Zapirain, and N. Nikolov, *The rank gradient from a combinatorial viewpoint*, Groups Geom. Dyn. **5** (2011), no. 2, 213–230, DOI 10.4171/GGD/124. MR2782170

[4] M. Abért and N. Nikolov, *Rank gradient, cost of groups and the rank versus Heegaard genus problem*, J. Eur. Math. Soc. (JEMS) **14** (2012), no. 5, 1657–1677, DOI 10.4171/JEMS/344. MR2966663

[5] I. Agol, *Criteria for virtual fibering*, J. Topol. **1** (2008), no. 2, 269–284, DOI 10.1112/jtopol/jtn003. MR2399130

[6] M. Albanese, L. Di Cerbo, and L. Lombardi, *Aspherical complex surfaces, the Singer Conjecture, and Gromov-lück inequality $\chi \geq |\sigma|$*, Preprint, `arXiv:2311.10226 [math.DG]`, 2023.

[7] N. Andrew, Y. Guerch, S. Hughes, and M. Kudlinska, *Homology growth of polynomially growing mapping tori*, Preprint, `arXiv:2305.10410`, 2023.

[8] N. Andrew, S. Hughes, and M. Kudlinska, *Torsion homology growth of polynomially growing free-by-cyclic groups*, Preprint, `arXiv:2211.04389`, 2022.

[9] M. F. Atiyah, *Elliptic operators, discrete groups and von Neumann algebras*, Colloque "Analyse et Topologie" en l'Honneur de Henri Cartan (Orsay, 1974), Astérisque, No. 32-33, Soc. Math. France, Paris, 1976, pp. 43–72. MR420729

[10] T. Austin, *Rational group ring elements with kernels having irrational dimension*, Proc. Lond. Math. Soc. (3) **107** (2013), no. 6, 1424–1448, DOI 10.1112/plms/pdt029. MR3149852

[11] G. Avramidi, *Rational manifold models for duality groups*, Geom. Funct. Anal. **28** (2018), no. 4, 965–994, DOI 10.1007/s00039-018-0449-8. MR3820436

[12] G. Avramidi, M. W. Davis, B. Okun, and K. Schreve, *The action dimension of right-angled Artin groups*, Bull. Lond. Math. Soc. **48** (2016), no. 1, 115–126, DOI 10.1112/blms/bdv083. MR3455755

[13] G. Avramidi, B. Okun, and K. Schreve, *Mod p and torsion homology growth in nonpositive curvature*, Invent. Math. **226** (2021), no. 3, 711–723, DOI 10.1007/s00222-021-01057-x. MR4337971

[14] G. Avramidi, B. Okun, and K. Schreve, *Homology growth, hyperbolization, and fibering*, Geom. Funct. Anal. **34** (2024), no. 2, 303–376, DOI 10.1007/s00039-024-00667-w. MR4715365

[15] W. Ballmann and J. Brüning, *On the spectral theory of manifolds with cusps*, J. Math. Pures Appl. (9) **80** (2001), no. 6, 593–625, DOI 10.1016/S0021-7824(00)01202-2. MR1842292

[16] N. Bergeron, P. Linnell, W. Lück, and R. Sauer, *On the growth of Betti numbers in p-adic analytic towers*, Groups Geom. Dyn. **8** (2014), no. 2, 311–329, DOI 10.4171/GGD/227. MR3231217

[17] N. Bergeron and A. Venkatesh, *The asymptotic growth of torsion homology for arithmetic groups*, J. Inst. Math. Jussieu **12** (2013), no. 2, 391–447, DOI 10.1017/S1474748012000667. MR3028790

[18] M. Bestvina, M. Kapovich, and B. Kleiner, *Van Kampen's embedding obstruction for discrete groups*, Invent. Math. **150** (2002), no. 2, 219–235, DOI 10.1007/s00222-002-0246-7. MR1933584

[19] J. Boschheidgen and A. Jaikin-Zapirain, *Twisted L^2-Betti numbers for sofic groups*, Bull. Lond. Math. Soc. **56** (2024), no. 6, 2178–2187. MR4768828

[20] M. R. Bridson and D. H. Kochloukova, *Volume gradients and homology in towers of residually-free groups*, Math. Ann. **367** (2017), no. 3-4, 1007–1045, DOI 10.1007/s00208-016-1387-0. MR3623217

[21] D. Burghelea, L. Friedlander, T. Kappeler, and P. McDonald, *Analytic and Reidemeister torsion for representations in finite type Hilbert modules*, Geom. Funct. Anal. **6** (1996), no. 5, 751–859, DOI 10.1007/BF02246786. MR1415762

[22] F. Calegari and M. Emerton, *Bounds for multiplicities of unitary representations of cohomological type in spaces of cusp forms*, Ann. of Math. (2) **170** (2009), no. 3, 1437–1446, DOI 10.4007/annals.2009.170.1437. MR2600878

[23] F. Calegari and M. Emerton, *Mod-p cohomology growth in p-adic analytic towers of 3-manifolds*, Groups Geom. Dyn. **5** (2011), no. 2, 355–366, DOI 10.4171/GGD/131. MR2782177

[24] J. Cheeger and M. Gromov, *L_2-cohomology and group cohomology*, Topology **25** (1986), no. 2, 189–215, DOI 10.1016/0040-9383(86)90039-X. MR837621

[25] M. Clay, *ℓ^2-torsion of free-by-cyclic groups*, Q. J. Math. **68** (2017), no. 2, 617–634, DOI 10.1093/qmath/haw057. MR3667215

[26] M. W. Davis and B. Okun, *Vanishing theorems and conjectures for the ℓ^2-homology of right-angled Coxeter groups*, Geom. Topol. **5** (2001), 7–74, DOI 10.2140/gt.2001.5.7. MR1812434

[27] J. Dodziuk, P. Linnell, V. Mathai, T. Schick, and S. Yates, *Approximating L^2-invariants and the Atiyah conjecture*, Comm. Pure Appl. Math. **56** (2003), no. 7, 839–873, DOI 10.1002/cpa.10076. Dedicated to the memory of Jürgen K. Moser. MR1990479

[28] J. Dubois, S. Friedl, and W. Lück, *The L^2-Alexander torsion of 3-manifolds*, J. Topol. **9** (2016), no. 3, 889–926, DOI 10.1112/jtopol/jtw013. MR3551842

[29] G. Elek and E. Szabó, *Hyperlinearity, essentially free actions and L^2-invariants. The sofic property*, Math. Ann. **332** (2005), no. 2, 421–441, DOI 10.1007/s00208-005-0640-8. MR2178069

[30] G. Elek and E. Szabó, *On sofic groups*, J. Group Theory **9** (2006), no. 2, 161–171, DOI 10.1515/JGT.2006.011. MR2220572

[31] M. Ershov and W. Lück, *The first L^2-Betti number and approximation in arbitrary characteristic*, Doc. Math. **19** (2014), 313–332. MR3178246

[32] M. Farber and S. Weinberger, *On the zero-in-the-spectrum conjecture*, Ann. of Math. (2) **154** (2001), no. 1, 139–154, DOI 10.2307/3062113. MR1847591

[33] D. R. Farkas and P. A. Linnell, *Congruence subgroups and the Atiyah conjecture*, Groups, rings and algebras, Contemp. Math., vol. 420, Amer. Math. Soc., Providence, RI, 2006, pp. 89–102, DOI 10.1090/conm/420/07970. MR2279234

[34] S. Friedl and W. Lück, *Universal L^2-torsion, polytopes and applications to 3-manifolds*, Proc. Lond. Math. Soc. (3) **114** (2017), no. 6, 1114–1151, DOI 10.1112/plms.12035. MR3661347

[35] S. Friedl and W. Lück, *L^2-Euler characteristics and the Thurston norm*, Proc. Lond. Math. Soc. (3) **118** (2019), no. 4, 857–900, DOI 10.1112/plms.12202. MR3938714

[36] S. Friedl and W. Lück, *The L^2-torsion function and the Thurston norm of 3-manifolds*, Comment. Math. Helv. **94** (2019), no. 1, 21–52, DOI 10.4171/CMH/453. MR3941465

[37] R. Frigerio, C. Löh, C. Pagliantini, and R. Sauer, *Integral foliated simplicial volume of aspherical manifolds*, Israel J. Math. **216** (2016), no. 2, 707–751, DOI 10.1007/s11856-016-1425-3. MR3557463

[38] F. Funke and D. Kielak, *Alexander and Thurston norms, and the Bieri-Neumann-Strebel invariants for free-by-cyclic groups*, Geom. Topol. **22** (2018), no. 5, 2647–2696, DOI 10.2140/gt.2018.22.2647. MR3811767

[39] D. Gaboriau, *Coût des relations d'équivalence et des groupes* (French, with English summary), Invent. Math. **139** (2000), no. 1, 41–98, DOI 10.1007/s002229900019. MR1728876

[40] D. Gaboriau, *Invariants l^2 de relations d'équivalence et de groupes* (French), Publ. Math. Inst. Hautes Études Sci. **95** (2002), 93–150, DOI 10.1007/s102400200002. MR1953191

[41] D. Gaboriau, *On orbit equivalence of measure preserving actions*, Rigidity in dynamics and geometry (Cambridge, 2000), Springer, Berlin, 2002, pp. 167–186. MR1919400

[42] D. Gaboriau and B. Seward, *Cost, ℓ^2-Betti numbers and the sofic entropy of some algebraic actions*, J. Anal. Math. **139** (2019), no. 1, 1–65, DOI 10.1007/s11854-020-0072-0. MR4041094

[43] Ł. Grabowski, *On Turing dynamical systems and the Atiyah problem*, Invent. Math. **198** (2014), no. 1, 27–69, DOI 10.1007/s00222-013-0497-5. MR3260857

[44] M. Gromov, *Volume and bounded cohomology*, Inst. Hautes Études Sci. Publ. Math. **56** (1982), 5–99 (1983). MR686042

[45] M. Gromov, *Large Riemannian manifolds*, Curvature and topology of Riemannian manifolds (Katata, 1985), Lecture Notes in Math., vol. 1201, Springer, Berlin, 1986, pp. 108–121, DOI 10.1007/BFb0075649. MR859578

[46] M. Gromov, *Hyperbolic groups*, Essays in group theory, Math. Sci. Res. Inst. Publ., vol. 8, Springer, New York, 1987, pp. 75–263, DOI 10.1007/978-1-4613-9586-7_3. MR919829

[47] M. Gromov, *Kähler hyperbolicity and L_2-Hodge theory*, J. Differential Geom. **33** (1991), no. 1, 263–292. MR1085144

[48] M. Gromov, *Asymptotic invariants of infinite groups*, Geometric group theory, Vol. 2 (Sussex, 1991), London Math. Soc. Lecture Note Ser., vol. 182, Cambridge Univ. Press, Cambridge, 1993, pp. 1–295. MR1253544

[49] M. Gromov, *Metric structures for Riemannian and non-Riemannian spaces*, Progress in Mathematics, vol. 152, Birkhäuser Boston, Inc., Boston, MA, 1999. Based on the 1981 French original [MR0682063 (85e:53051)]; With appendices by M. Katz, P. Pansu and S. Semmes; Translated from the French by Sean Michael Bates. MR1699320

[50] F. Henneke, *Embeddings of group rings and L^2-invariants*, PhD Thesis, Bonn, 2021.

[51] F. Henneke and D. Kielak, *Agrarian and L^2-invariants*, Fund. Math. **255** (2021), no. 3, 255–287, DOI 10.4064/fm808-4-2021. MR4324826

[52] E. Hess and T. Schick, *L^2-torsion of hyperbolic manifolds*, Manuscripta Math. **97** (1998), no. 3, 329–334, DOI 10.1007/s002290050105. MR1654784

[53] F. Hirzebruch, *The signature theorem: reminiscences and recreation*, Prospects in mathematics (Proc. Sympos., Princeton Univ., Princeton, N.J., 1970), Ann. of Math. Stud., No. 70, Princeton Univ. Press, Princeton, NJ, 1971, pp. 3–31. MR368023

[54] N. V. Ivanov, *Foundations of the theory of bounded cohomology* (Russian, with English summary), Zap. Nauchn. Sem. Leningrad. Otdel. Mat. Inst. Steklov. (LOMI) **143** (1985), 69–109, 177–178. Studies in topology, V. MR806562

[55] A. Jaikin-Zapirain, *The base change in the Atiyah and the Lück approximation conjectures*, Geom. Funct. Anal. **29** (2019), no. 2, 464–538, DOI 10.1007/s00039-019-00487-3. MR3945838

[56] A. Jaikin-Zapirain, *L^2-Betti numbers and their analogues in positive characteristic*, Groups St Andrews 2017 in Birmingham, London Math. Soc. Lecture Note Ser., vol. 455, Cambridge Univ. Press, Cambridge, 2019, pp. 346–405. MR3931420

[57] A. Jaikin-Zapirain, *The universality of Hughes-free division rings*, Selecta Math. (N.S.) **27** (2021), no. 4, Paper No. 74, 33, DOI 10.1007/s00029-021-00691-w. MR4292784

[58] A. Jaikin-Zapirain and D. López-Álvarez, *The strong Atiyah and Lück approximation conjectures for one-relator groups*, Math. Ann. **376** (2020), no. 3-4, 1741–1793, DOI 10.1007/s00208-019-01926-0. MR4081128

[59] J. Jost and Y. L. Xin, *Vanishing theorems for L^2-cohomology groups*, J. Reine Angew. Math. **525** (2000), 95–112, DOI 10.1515/crll.2000.071. MR1780428

[60] H. Kammeyer, *Introduction to ℓ^2-invariants*, Lecture Notes in Mathematics, vol. 2247, Springer, Cham, 2019, DOI 10.1007/978-3-030-28297-4. MR3971279

[61] H. Kammeyer and S. Kionke, *On the profinite rigidity of lattices in higher rank Lie groups*, Math. Proc. Cambridge Philos. Soc. **174** (2023), no. 2, 369–384, DOI 10.1017/S0305004122000305. MR4545210

[62] D. Kielak, *The Bieri-Neumann-Strebel invariants via Newton polytopes*, Invent. Math. **219** (2020), no. 3, 1009–1068, DOI 10.1007/s00222-019-00919-9. MR4055183

[63] D. Kielak, *Residually finite rationally solvable groups and virtual fibring*, J. Amer. Math. Soc. **33** (2020), no. 2, 451–486, DOI 10.1090/jams/936. MR4073866

[64] D. Kielak and M. Linton, *Group rings of three-manifold groups*, Proc. Amer. Math. Soc. **152** (2024), no. 5, 1939–1946, DOI 10.1090/proc/16716. MR4728464

[65] D. Kielak and B. Sun, *Agrarian and L^2-Betti numbers of locally indicable groups, with a twist*, Preprint, `arXiv:2112.07394`, 2021.

[66] B. Kleiner and J. Lott, *Notes on Perelman's papers*, Geom. Topol. **12** (2008), no. 5, 2587–2855, DOI 10.2140/gt.2008.12.2587. MR2460872

[67] A. Knebusch, P. Linnell, and T. Schick, *On the center-valued Atiyah conjecture for L^2-Betti numbers*, Doc. Math. **22** (2017), 659–677. MR3650226

[68] M. Lackenby, *Expanders, rank and graphs of groups*, Israel J. Math. **146** (2005), 357–370, DOI 10.1007/BF02773541. MR2151608

[69] M. Lackenby, *Large groups, property (τ) and the homology growth of subgroups*, Math. Proc. Cambridge Philos. Soc. **146** (2009), no. 3, 625–648, DOI 10.1017/S0305004108002089. MR2496348

[70] M. Lackenby, *Detecting large groups*, J. Algebra **324** (2010), no. 10, 2636–2657, DOI 10.1016/j.jalgebra.2010.07.012. MR2725193

[71] T. Le, *Homology torsion growth and Mahler measure*, Comment. Math. Helv. **89** (2014), no. 3, 719–757, DOI 10.4171/CMH/332. MR3260847

[72] H. Li and A. Thom, *Entropy, determinants, and L^2-torsion*, J. Amer. Math. Soc. **27** (2014), no. 1, 239–292, DOI 10.1090/S0894-0347-2013-00778-X. MR3110799

[73] W. Lück and P. Linnell, *Localization, Whitehead groups and the Atiyah conjecture*, Ann. K-Theory **3** (2018), no. 1, 33–53, DOI 10.2140/akt.2018.3.33. MR3695363

[74] P. Linnell, W. Lück, and R. Sauer, *The limit of $\mathbb{F}_p$-Betti numbers of a tower of finite covers with amenable fundamental groups*, Proc. Amer. Math. Soc. **139** (2011), no. 2, 421–434, DOI 10.1090/S0002-9939-2010-10689-5. MR2736326

[75] P. Linnell, B. Okun, and T. Schick, *The strong Atiyah conjecture for right-angled Artin and Coxeter groups*, Geom. Dedicata **158** (2012), 261–266, DOI 10.1007/s10711-011-9631-y. MR2922714

[76] P. Linnell and T. Schick, *Finite group extensions and the Atiyah conjecture*, J. Amer. Math. Soc. **20** (2007), no. 4, 1003–1051, DOI 10.1090/S0894-0347-07-00561-9. MR2328714

[77] P. A. Linnell, *Division rings and group von Neumann algebras*, Forum Math. **5** (1993), no. 6, 561–576, DOI 10.1515/form.1993.5.561. MR1242889

[78] Y. Liu, *Degree of L^2-Alexander torsion for 3-manifolds*, Invent. Math. **207** (2017), no. 3, 981–1030, DOI 10.1007/s00222-016-0680-6. MR3608287

[79] C. Löh and M. Uschold, *L^2-Betti numbers and computability of reals*, Computability **12** (2023), no. 2, 175–201. MR4636004

[80] J. Lott, *Heat kernels on covering spaces and topological invariants*, J. Differential Geom. **35** (1992), no. 2, 471–510. MR1158345

[81] J. Lott, *The zero-in-the-spectrum question*, Enseign. Math. (2) **42** (1996), no. 3-4, 341–376. MR1426443

[82] W. Lück, *Approximating L^2-invariants by their finite-dimensional analogues*, Geom. Funct. Anal. **4** (1994), no. 4, 455–481, DOI 10.1007/BF01896404. MR1280122

[83] W. Lück, *L^2-Betti numbers of mapping tori and groups*, Topology **33** (1994), no. 2, 203–214, DOI 10.1016/0040-9383(94)90011-6. MR1273782

[84] W. Lück, *L^2-torsion and 3-manifolds*, Low-dimensional topology (Knoxville, TN, 1992), Conf. Proc. Lecture Notes Geom. Topology, III, Int. Press, Cambridge, MA, 1994, pp. 75–107. MR1316175

[85] W. Lück, *Dimension theory of arbitrary modules over finite von Neumann algebras and L^2-Betti numbers. I. Foundations*, J. Reine Angew. Math. **495** (1998), 135–162, DOI 10.1515/crll.1998.015. MR1603853

[86] W. Lück, *Dimension theory of arbitrary modules over finite von Neumann algebras and L^2-Betti numbers. II. Applications to Grothendieck groups, L^2-Euler characteristics and Burnside groups*, J. Reine Angew. Math. **496** (1998), 213–236, DOI 10.1515/crll.1998.031. MR1605818

[87] W. Lück, *L^2-invariants: theory and applications to geometry and K-theory*, Ergebnisse der Mathematik und ihrer Grenzgebiete. 3. Folge. A Series of Modern Surveys in Mathematics [Results in Mathematics and Related Areas. 3rd Series. A Series of Modern Surveys in Mathematics], vol. 44, Springer-Verlag, Berlin, 2002, DOI 10.1007/978-3-662-04687-6. MR1926649

[88] W. Lück, *Survey on classifying spaces for families of subgroups*, Infinite groups: geometric, combinatorial and dynamical aspects, Progr. Math., vol. 248, Birkhäuser, Basel, 2005, pp. 269–322, DOI 10.1007/3-7643-7447-0_7. MR2195456

[89] W. Lück, *Approximating L^2-invariants and homology growth*, Geom. Funct. Anal. **23** (2013), no. 2, 622–663, DOI 10.1007/s00039-013-0218-7. MR3053758

[90] W. Lück, *Approximating L^2-invariants by their classical counterparts*, EMS Surv. Math. Sci. **3** (2016), no. 2, 269–344, DOI 10.4171/EMSS/18. MR3576534

[91] W. Lück, *Twisting L^2-invariants with finite-dimensional representations*, J. Topol. Anal. **10** (2018), no. 4, 723–816, DOI 10.1142/S1793525318500279. MR3881040

[92] W. Lück, *Survey on L^2-invariants and 3-manifolds*, Bull. Lond. Math. Soc. **53** (2021), no. 6, 1583–1620, DOI 10.1112/blms.12536. MR4368687

[93] W. Lück, *Isomorphism Conjectures in K- and L-theory*, in preparation, see `http://him-lueck.uni-bonn.de/data/ic.pdf`, 2024.

[94] W. Lück and D. Osin, *Approximating the first L^2-Betti number of residually finite groups*, J. Topol. Anal. **3** (2011), no. 2, 153–160, DOI 10.1142/S1793525311000532. MR2819192

[95] W. Lück and M. Rothenberg, *Reidemeister torsion and the K-theory of von Neumann algebras*, K-Theory **5** (1991), no. 3, 213–264, DOI 10.1007/BF00533588. MR1162441

[96] W. Lück, R. Sauer, and C. Wegner, *L^2-torsion, the measure-theoretic determinant conjecture, and uniform measure equivalence*, J. Topol. Anal. **2** (2010), no. 2, 145–171, DOI 10.1142/S179352531000032X. MR2652905

[97] W. Lück and T. Schick, *L^2-torsion of hyperbolic manifolds of finite volume*, Geom. Funct. Anal. **9** (1999), no. 3, 518–567, DOI 10.1007/s000390050095. MR1708444

[98] W. Lück and T. Schick, *Various L^2-signatures and a topological L^2-signature theorem*, High-dimensional manifold topology, World Sci. Publ., River Edge, NJ, 2003, pp. 362–399, DOI 10.1142/9789812704443_0015. MR2048728

[99] W. Lück and T. Schick, *Approximating L^2-signatures by their compact analogues*, Forum Math. **17** (2005), no. 1, 31–65, DOI 10.1515/form.2005.17.1.31. MR2110537

[100] V. Mathai, *L^2-analytic torsion*, J. Funct. Anal. **107** (1992), no. 2, 369–386, DOI 10.1016/0022-1236(92)90114-X. MR1172031

[101] J. Morgan and G. Tian, *The geometrization conjecture*, Clay Mathematics Monographs, vol. 5, American Mathematical Society, Providence, RI; Clay Mathematics Institute, Cambridge, MA, 2014. MR3186136

[102] B. Okun and K. Schreve, *The L^2-(co)homology of groups with hierarchies*, Algebr. Geom. Topol. **16** (2016), no. 5, 2549–2569, DOI 10.2140/agt.2016.16.2549. MR3572340

[103] M. Olbrich, *L^2-invariants of locally symmetric spaces*, Doc. Math. **7** (2002), 219–237. MR1938121

[104] D. Osin, *Rank gradient and torsion groups*, Bull. Lond. Math. Soc. **43** (2011), no. 1, 10–16, DOI 10.1112/blms/bdq075. MR2765544

[105] V. G. Pestov, *Hyperlinear and sofic groups: a brief guide*, Bull. Symbolic Logic **14** (2008), no. 4, 449–480, DOI 10.2178/bsl/1231081461. MR2460675

[106] S. Popa, *Deformation and rigidity for group actions and von Neumann algebras*, International Congress of Mathematicians. Vol. I, Eur. Math. Soc., Zürich, 2007, pp. 445–477, DOI 10.4171/022-1/18. MR2334200

[107] J. Raimbault, *Exponential growth of torsion in abelian coverings*, Algebr. Geom. Topol. **12** (2012), no. 3, 1331–1372, DOI 10.2140/agt.2012.12.1331. MR2966689

[108] A. A. Ranicki, *Algebraic L-theory and topological manifolds*, Cambridge Tracts in Mathematics, vol. 102, Cambridge University Press, Cambridge, 1992. MR1211640

[109] D. B. Ray and I. M. Singer, *R-torsion and the Laplacian on Riemannian manifolds*, Advances in Math. **7** (1971), 145–210, DOI 10.1016/0001-8708(71)90045-4. MR295381

[110] H. Reich, *L^2-Betti numbers, isomorphism conjectures and noncommutative localization*, Non-commutative localization in algebra and topology, London Math. Soc. Lecture Note Ser., vol. 330, Cambridge Univ. Press, Cambridge, 2006, pp. 103–142, DOI 10.1017/CBO9780511526381.013. MR2222484

[111] P. Sánchez-Peralta, *On vanishing criteria of L^2-Betti numbers of groups*, Preprint, `arXiv:2307.07031`, 2023.

[112] R. Sauer, *Amenable covers, volume and L^2-Betti numbers of aspherical manifolds*, J. Reine Angew. Math. **636** (2009), 47–92, DOI 10.1515/CRELLE.2009.082. MR2572246

[113] R. Sauer, *Volume and homology growth of aspherical manifolds*, Geom. Topol. **20** (2016), no. 2, 1035–1059, DOI 10.2140/gt.2016.20.1035. MR3493098

[114] R. Sauer and A. Thom, *A spectral sequence to compute L^2-Betti numbers of groups and groupoids*, J. Lond. Math. Soc. (2) **81** (2010), no. 3, 747–773, DOI 10.1112/jlms/jdq017. MR2650795

[115] J. A. Schafer, *Topological Pontrjagin classes*, Comment. Math. Helv. **45** (1970), 315–332, DOI 10.1007/BF02567335. MR275443

[116] T. Schick, *L^2-determinant class and approximation of L^2-Betti numbers*, Trans. Amer. Math. Soc. **353** (2001), no. 8, 3247–3265, DOI 10.1090/S0002-9947-01-02699-X. MR1828605

[117] J.-C. Schlage-Puchta, *A p-group with positive rank gradient*, J. Group Theory **15** (2012), no. 2, 261–270, DOI 10.1515/jgt.2011.101. MR2900227

[118] K. Schmidt, *Dynamical systems of algebraic origin*, Progress in Mathematics, vol. 128, Birkhäuser Verlag, Basel, 1995. MR1345152

[119] M. Schmidt, *L^2-Betti Numbers of $\mathcal{R}$-Spaces and the Integral Foliated Simplicial Volume*, PhD Thesis, WWU Münster, 2005. Available online at `http://nbn-resolving.de/urn:nbn:de:hbz:6-05699458563`.

[120] P. Scholze and D. Clausen, *Lectures on analytic geometry*, lecture notes URL: `https://people.mpim-bonn.mpg.de/scholze/Analytic.pdf`, 2019.

[121] P. Scholze and D. Clausen, *Lectures on condensed mathematics*, lecture notes URL: `https://people.mpim-bonn.mpg.de/scholze/Condensed.pdf`, 2019.

[122] P. Scholze and D. Clausen, *Condensed mathematics and complex geometry*, lecture notes URL: `https://people.mpim-bonn.mpg.de/scholze/Complex.pdf`, 2022.

[123] T. N. Shorey and R. Tijdeman, *Exponential Diophantine equations*, Cambridge Tracts in Mathematics, vol. 87, Cambridge University Press, Cambridge, 1986, DOI 10.1017/CBO9780511566042. MR891406

[124] M. Uschold, *Torsion homology growth and cheap rebuilding of inner-amenable groups*, Preprint, `arXiv:2212.07916`, 2022.

[125] C. T. C. Wall, *Poincaré complexes. I*, Ann. of Math. (2) **86** (1967), 213–245, DOI 10.2307/1970688. MR217791

Mathematical Institute of the University of Bonn, Endenicher Allee 60, 53115 Bonn, Germany

Email address: `kirstein@math.uni-bonn.de`
URL: `https://dkirstein.github.io`

Mathematical Institute of the University of Bonn, Endenicher Allee 60, 53115 Bonn, Germany

Email address: `kremer@mpim-bonn.mpg.de`
URL: `https://sites.google.com/view/christian-kremer-math`

Mathematical Institute of the University of Bonn, Endenicher Allee 60, 53115 Bonn, Germany

Email address: `wolfgang.lueck@him.uni-bonn.de`
URL: `http://www.him.uni-bonn.de/lueck`

Contemporary Mathematics
Volume **816**, 2025
https://doi.org/10.1090/conm/816/16350

Finiteness properties of algebraic fibers of group extensions

Dessislava H. Kochloukova and Stefano Vidussi,
with an appendix by Marco Boggi

ABSTRACT. This survey describes some recent work, by the authors and others, on the existence of algebraic fibrations of group extensions, as well as the finiteness properties of their algebraic fibers, in the realm of both abstract and pro-p groups. We also discuss some applications of these results to (higher) coherence.

1. Introduction

Let G be a finitely generated group, and let $\phi\colon G \to \mathbb{Z}$ be an epimorphism. We say that ϕ is an *algebraic fibration* if $\operatorname{Ker}\phi \trianglelefteq G$ is finitely generated (f.g.), in which case we refer to $\operatorname{Ker}\phi$ as an *algebraic fiber*. The study of algebraic fibrations of groups is a storied and, at the same time, current topic of research in group theory. It dates back, at least morally, to the study of *fibered knots*, classical knots whose exterior $N(K) = S^3 \setminus \nu K$ is a locally trivial surface bundle over a circle, developed (among others) by Lee Neuwirth, Elvira Rapaport Strasser, and John Stallings in the early 1960s. The surface in object is compact, with a single boundary component, hence its fundamental group is F_{2g}, the free group on $2g$ generators, where g is the genus of the knot. (Some of the facts casually stated here are actually nontrivial.) The long exact sequence of a fibration reduces, in this case, to the fact that the abelianization map $\phi\colon \pi_1(N(K)) \to \mathbb{Z}$ has kernel F_{2g}. More generally, given a compact, irreducible (i.e., such that all 2-spheres bound 3-disks) oriented 3-manifold N that fibers over the circle (which entails that the boundary, if nonempty, is a disjoint union of tori), we have an epimorphism $\phi\colon \pi_1(N) \to \mathbb{Z}$ with kernel a surface group, i.e., an algebraic fibration. In the 3-manifold case, a remarkable fact, due to Stallings, is that an algebraic fibration on a 3-manifold group $\pi_1(N)$ actually "integrates" to a locally trivial surface bundle structure on N, and in particular the group $\operatorname{Ker}\phi$ is the fundamental group of a finite complex. This fact is simultaneously central and peculiar to the study of 3-manifolds.

2020 *Mathematics Subject Classification.* Primary 20J05, 57M07, 14F35.

The first author was partially supported by Bolsa de produtividade em pesquisa CNPq 305457/2021-7 and Projeto temático FAPESP 18/23690-6.

The realm of this survey is the class of groups that can be written as extensions of f.g. groups, i.e., appear in a short exact sequence of the type

$$1 \longrightarrow K \longrightarrow G \xrightarrow{f} \Gamma \longrightarrow 1 \tag{1}$$

where K and Γ are themselves finitely generated. With a slight abuse of notation, we will refer to groups of this type *tout court* as extensions. Our goal is to determine conditions under which these extensions admit a algebraic fibrations and discuss, in a sense, "what can go wrong" (compared with the 3-manifold case) in terms of regularity of the fiber. In particular, we will find instances of a variety of finiteness properties that the kernels of algebraic fibrations (even in geometric situations) can have.

This survey is structured as follows. In Section 2 we will define the finiteness conditions we are interested in and their interplay with the BNSR invariants. In Section 3 we will introduce the class of groups in the title, namely groups that are extensions of a f.g. group by a f.g. group. Section 4 contains the main results implying or obstructing finiteness of the algebraic fibers, as well as some applications. Section 5 discusses pro-p groups, their presentations, homology, and finiteness. Section 6 relates the theories of finitely presented pro-p and (abstract) metabelian groups. Section 7 discusses the pro-p version of some of the main results in Section 4, and Section 8 is devoted to some applications of these results to coherence. Further results on coherence follow by applying the results contained in the Appendix written by Marco Boggi.

2. Finiteness properties of groups and BNSR invariants

We start by introducing finiteness properties of groups, that can be thought of as strengthening of the condition of finite generation. These come in two broad flavors, that can be thought of as *homotopical* or *homological.* (We refer [**Br94**, **Br10**] for a thorough discussion of these notions, and the many related facts that will be implicitly used in what follows.)

DEFINITION 2.1. Let G be a group and let n be a nonnegative integer.

- we say that G *has type* F_n if it admits as classifying space a CW-complex $K(G,1)$, which we can assume without loss of generality to have a single vertex, whose n-skeleton is finite; furthermore we say it has type F if the $K(G,1)$ is a finite complex;
- we say G *has type* $FP_n(R)$, where R is a commutative ring with unity, if R, as a trivial $R[G]$-module, admits a resolution by projective $R[G]$-modules

  $$\ldots \longrightarrow P_i \longrightarrow P_{i-1} \ldots \longrightarrow P_0 \longrightarrow R \longrightarrow 0,$$

  with the first n-terms being f.g. as $R[G]$-modules.

We usually write FP_n to mean type $FP_n(\mathbb{Z})$. For a group G, finiteness type F_1 (or $FP_1(R)$) is equivalent to being f.g., and finiteness type F_2 is equivalent to finite presentation (f.p. henceforth); thanks to work of Bestvina and Brady there exist however groups of type FP_2 (classically referred to as *almost finitely presented*), and even FP_∞, that fail to be F_2; see [**BB97**]. In more generality, finiteness type F_n entails $FP_n(R)$ for all R, but the converse fails. In Section 5 we will extend this definition to the case of pro-p groups.

As outlined before, we will mostly be concerned with the finiteness properties of algebraic fibers, which are by definition kernels of an epimorphism from a group

G to $\mathbb{Z}$, hence in particular $[G, G] \trianglelefteq \operatorname{Ker} \phi \trianglelefteq G$. Conveniently, the study of the finiteness properties of such subgroups fits in the broader context of the Bieri–Neumann–Strebel–Renz (BNSR) invariants of the "ambient" group G. Much like finiteness properties, these exist in both homotopical and homological form. These were defined originally in [**BNS87**] (for the $n = 1$ case in which they are more often referred to as BNS invariants), and then extended to all n in [**R88, BR88**].

In order to define them, consider the projectivization of the first cohomology group $H^1(G;\mathbb{R})$, namely the *character sphere*

$$S(G) := (H^1(G;\mathbb{R}) \setminus \{0\})/\mathbb{R}^+;$$

elements $\chi \in S(G)$ are referred to as *characters*, and a character is *discrete* if one of its representatives in $H^1(G;\mathbb{R})$ is contained in the lattice $H^1(G;\mathbb{Z}) \subset H^1(G;\mathbb{R})$. Stated differently, there exists a representative of χ which is given by an epimorphism $\phi\colon G \to \mathbb{Z}$.

It is convenient to define the 0th homotopical and homological invariant $\Sigma^0(G) = \Sigma^0(G, R) = S(G)$. To define the nth (homotopical) BNSR invariant $\Sigma^n(G) \subset S(G)$, assume now that G has type F_n, and let Γ_n be the n-skeleton of the universal cover of some $K(G, 1)$. Choosing $K(G, 1)$ to have a single vertex, Γ_0 is in bijection with G and Γ_1 is the Cayley graph for some finite generating set. Given a character $\chi \in S(G)$, let $(\Gamma_n)_\chi$ be the subcomplex of Γ_n defined by the vertices $v \in \Gamma_0$ that satisfy $\chi(v) \geq 0$ (this definition is independent of the choice of representative of $S(G)$ in $H^1(G;\mathbb{R})$).

$$\Sigma^n(G) = \{\chi \in S(G) | (\Gamma_n)_\chi \text{ is } (n-1)\text{-connected for some choice of } \Gamma_n\}.$$

To define the nth (homological) BNSR invariant $\Sigma^n(G, R) \subset S(G)$, define much as above G_χ to be the submonoid defined by the elements $g \in G$ such that $\chi(g) \geq 0$. We say that R is of type FP_n as a (trivial) $R[G_\chi]$ module if it admits a resolution with the first n-terms f.g. projective $R[G_\chi]$-modules. Then

$$\Sigma^n(G, R) = \{\chi \in S(G) | R \text{ is of type } FP_n \text{ as } R[G_\chi]\text{-module}\}.$$

If G is not of type $FP_n(R)$, we put $\Sigma^n(G; R) = \emptyset$.

The first BNSR invariant $\Sigma^1(G) = \Sigma^1(G, R)$ is the same for the homotopical and homological variants, and irrespective of the choice of ring R, while the invariants may differ at higher degree.

The connection between the finiteness properties of algebraic fibers and BNSR invariants is distilled in the following result, whose proof combines work contained in [**BR88, R88, R89**].

PROPOSITION 2.1 ([**BR88, R88, R89**]). *Let G be a f.g. group, and let $\chi \in S(G)$ be a discrete character. If G has type F_n (respectively, type FP_n), then* $\operatorname{Ker} \chi \trianglelefteq G$ *has type F_n (respectively, type FP_n if and only if $\pm\chi \in \Sigma^n(G)$ (respectively, $\pm\chi \in \Sigma^n(G, R)$).*

On face value, this proposition embeds the problem of determining the finiteness properties of an algebraic fiber of G into the broader problem of studying the BNSR invariants of G: this brings an entire apparatus to the study of the original problem, and in particular offers the advantage that the BNSR invariants of G are *open* subsets of the character sphere (endowed with the subspace topology arising as unit sphere $S(G) \subset H^1(G;\mathbb{R}) \cong \mathbb{R}^{b_1(G)}$). The added flexibility of working with nondiscrete characters is comparable to using cohomology with real coefficients as opposed to integer ones, and this is frequently more than analogy and becomes a guiding principle, as in the development of the Thurston norm as tool to study genus-minimizing surfaces in 3-manifolds.

3. Extensions

The study of BNSR invariants of groups is tightly related with our ability to understand the properties of the group, and it is rarely straightforward. The scope of this survey is therefore quite narrow, and is directed towards a very specific class of groups, namely f.g. groups G that arise as extensions of f.g. groups. Two classes of groups come to mind in that realm: free-by-free groups (where both K and Γ are free groups) and fundamental groups of aspherical manifolds, which admit a locally trivial fibration. The latter class notably contains surface-by-surface groups.

The case where the extension is trivial, i.e., $G = K \times \Gamma$, is the first example that comes to mind. In that case it is well known that the first BNSR invariant of G can be read directly from those of K and Γ: precisely, we have the relation (more easily phrased in terms of complementary subsets)

$$\Sigma^1(G)^c = \bigcup_{0 \leq p \leq 1} \Sigma^p(K)^c * \Sigma^{1-p}(\Gamma)^c$$

and a similar relationship (necessarily) holds for the homological invariants. (Here, characters on K and Γ are interpreted as characters on G by pull-back.) Extending this result to $n \geq 2$ is notoriously intricate and partial results, as well as counterexamples to the most obvious generalizations (the "product formula"), hold; see [**BG10**]. One could hope, at a minimum, that the BNSR invariants of G are "affected" by those of K and Γ, and in fact it is not difficult to show that if a character of Γ has a given finiteness property, so does the corresponding pull-back character of G. The situation is more complex regarding characters from K because they may not extend, or they do not canonically, as characters of G (there is no pull-back!). There is one easy exception (which, in a sense, is complementary to the product case), and that is the case where $b_1(G) = b_1(\Gamma)$, so that $S(G) = S(\Gamma)$ (properly, under identification under pull-back). In that case it is not too difficult to see that as we have assumed that K is f.g., then $\Sigma^1(G) = \Sigma^1(\Gamma)$. (Assuming suitable stronger assumptions on the finiteness properties of K, we can deduce a similar equality result for higher BNSR invariants.)

In fact, this result generalizes to a class of groups that can be written as extensions, and that arise in complex geometry, namely Kähler groups.

DEFINITION 3.1. A complex manifold X is Kähler if it can be endowed with an Hermitian metric such that the associated 2-form is closed. Fundamental groups of compact Kähler manifolds are referred to as *Kähler groups*.

Differently from what happens with complex manifolds (which can have any f.p. group as fundamental group) Kähler groups admit significant restrictions. Implicitly, one such restriction is the outcome of the next result, appearing in [**De10**], that constraints the form of the first BNS invariant for Kähler groups (obviously, with $b_1(G) > 0$). We will formulate it in a way that emphasizes the fact, *a priori* nonobvious, that these groups can be thought of as extensions as in Equation (1).

THEOREM 3.1 ([**De10**]). *Let G be a Kähler group. Then either*

- *G has finitely generated commutator subgroup, hence is an extension of the form*
$$1 \longrightarrow [G,G] \longrightarrow G \longrightarrow H_1(G) \longrightarrow 1$$
or
- *there exist a finite collection of hyperbolic orbisurface groups $\Gamma_i, i \in I$ and extensions*
$$1 \longrightarrow K_i \longrightarrow G \xrightarrow{f_i} \Gamma_i \longrightarrow 1$$
such that
$$\Sigma^1(G) = S(G) \setminus \coprod_{i\in I} f_i^* S(\Gamma_i)$$

Some remarks are in order. First, the reader can find the definition of hyperbolic orbisurface groups in [**BU11**] (which is an excellent survey on the result above), but for our intents it is sufficient to know that they contain as a finite index subgroup the fundamental group of an hyperbolic surface, which is known to have empty BNSR invariants. From this it follows that $\Sigma^1(\Gamma_i) = \emptyset$, which explains why all characters of G that pull back from the Γ_i are singular. What is not trivial is that all other characters are in $\Sigma^1(G)$. Furthermore, unless there exists an $f_i\colon G \longrightarrow \Gamma_i$ inducing an isomorphism in homology with rational coefficients (in which case we say that X, and by extension G, has *Albanese dimension 1* and f_i can be proven to be essentially unique), then $\Sigma^1(G)$ is nonempty and, by openness, G admits an algebraic fibration. Lastly, one can prove that the subspaces $f_i^*(H^1(\Gamma_i, \mathbb{Q})) \subset H^1(G;\mathbb{Q})$ are pairwise orthogonal; see, e.g., [**Ca03**]

A rather informative example of the application of Theorem 3.1 comes by looking at Kodaira fibrations. These are Kähler manifolds (more precisely, complex projective surfaces) which admit one (or more) structure of surface bundle over a surface in such a way that the fibration map is holomorphic. It follows that their fundamental group G is a surface-by-surface group, and barring the case where these manifolds have Albanese dimension 1, we have $\Sigma^1(G) \neq \emptyset$. This (and obviously much more) is a consequence from Theorem 3.1 and in our understanding a proof not involving the fact that these groups are Kähler did not exist.

4. Finiteness properties of the fibers

In light of Theorem 3.1 one can ask whether there is a proof that $\Sigma^1(G) \neq \emptyset$ for non-Kähler surface-by-surface groups. (For rather deep reasons, namely the Parshin–Arakelov rigidity, surface-by-surface bundles are "generically" not Kähler.) More generally, one can ask for necessary conditions for an extension as in Equation (1) to have nonempty higher BNSR invariants, using as tool only the fact that the group is an extension (as opposed to Theorem 3.1, whose proof hinges on the Kähler assumption). The first results in that sense are contained in [**KW22,FV23**], which deal with the first BNSR invariant (i.e., $n_0 = 1$ in the notation below), and these are extended in [**KV23**].

THEOREM 4.1 ([**KW22, FV23, KV23**]). *Let $1 \to K \to G \to \Gamma \to 1$ be a short exact sequence of groups such that G and K are of type F_{n_0} (resp., $FP_{n_0}(R)$) and $b_1(\Gamma) > 0$. Suppose that $\phi\colon K \to \mathbb{Z}$ is an epimorphism with kernel* $\operatorname{Ker}\phi = N$ *of type F_{n_0-1} (resp., $FP_{n_0-1}(R)$) such that ϕ extends to a real homomorphism of G. Then there exists an epimorphism $\psi\colon G \to \mathbb{Z}$ that extends ϕ such that* $\operatorname{Ker}\psi$ *is of* *type F_{n_0} (resp., $FP_{n_0}(R)$). Furthermore, if K, G, and N are of type F_∞ (resp., $FP_\infty(R)$) then $\operatorname{Ker}\psi$ can be chosen of type F_∞ (resp. $FP_\infty(R)$).*

REMARK. Note that when $n_0 = 1$, the existence of a $\phi\colon K \to \mathbb{Z}$ that extends to G is equivalent to the condition that the inclusion-induced homomorphism $H_1(K) \to H_G$ (which, by the Lyndon–Hochschild–Serre sequence factorizes through the quotient map $H_1(K) \to H_1(K)_\Gamma$) has image of positive rank, or, equivalently, $b_1(G) > b_1(\Gamma)$. This condition is referred to by saying that the extension in Equation (1) has excessive homology. As any $\phi\colon K \to \mathbb{Z}$ has kernel of type (at least) F_0, if the extension has excessive homology then G admits an epimorphism $\psi\colon G \to \mathbb{Z}$ with kernel of type F_1, hence $\pm[\psi] \in \Sigma^1(G)$.

To give an idea of the proof, start with a presentation of the base group Γ, thought of as an epimorphism from the free group F_n to Γ for suitable n. This epimorphism induces a diagram

$$\begin{array}{ccccccccc} 1 & \longrightarrow & K & \longrightarrow & \Pi & \longrightarrow & F_n & \longrightarrow & 1 \\ & & \downarrow\cong & & \downarrow & & \downarrow & & \\ 1 & \longrightarrow & K & \longrightarrow & G & \longrightarrow & \Gamma & \longrightarrow & 1 \end{array} \tag{2}$$

where all vertical maps are epimorphisms.

Observe that we can write Π as amalgamated product

$$\Pi = \Pi_1 *_K \Pi_2 *_K \cdots *_K \Pi_n$$

where each Π_i has the form of an HNN extension $\Pi_i = \langle K, s_i | s_i k s_i^{-1} = f_i(k)\rangle$ for some automorphism $f_i\colon K \to K$. We can then work with the group Π: the advantage of doing so is that amalgamated products are, in a sense, easier to deal with in terms of the study of the BNSR invariants. The condition on excessive homology allows one to establish the existence of algebraic fibrations on Π, by "perturbing" any of the fibrations of the Π_i. (To do so, the openness, and the very use, of the BNSR invariants, is essential.) As long as $b_1(\Gamma) > 0$ at least some of the fibrations descend to G, and with some delicate arguments one can control the finiteness properties of the fiber to the extent stated.

We want to discuss an interesting consequence of Theorem 4.1, which follows from repeated application of this result to iterated extension. Omitting, for the sake of simplicity, full generality (for which the reader is referred to [**KV23**, Theorem 1.6]) we will consider the following set up.

DEFINITION 4.1. A *normal filtration of length* n of a group G is a collection of normal subgroups $G_i \trianglelefteq G$, $0 \le i \le n$ such that

$$1 = G_0 \trianglelefteq G_1 \trianglelefteq \ldots \trianglelefteq G_{n-1} \trianglelefteq G_n = G.$$

We say that G is poly-free (respectively, poly-surface) if the factor groups $\Gamma_{j+1} := G_{j+1}/G_j$, $0 \le j \le n-1$ are f.g. free (respectively, surface groups).

REMARK. In literature the terms poly-free and poly-surface usually apply also to the case where the filtration is just *subnormal*, i.e., each element of the filtration is normal in the successor, but not necessarily in G. The result below applies to the more general case, but statements can become quite cumbersome, so we limit ourselves to the class above.

We have the following.

THEOREM 4.2 ([**KV23**]). *Let G be a poly-free or a poly-surface group of length n, and assume that for all $0 \leq j \leq n-1$ the sequence*

$$1 \longrightarrow G_{j+1}/G_j \longrightarrow G/G_j \longrightarrow G/G_{j+1} \longrightarrow 1 \tag{3}$$

has excessive homology. Then for every discrete character $\phi\colon G_1 \to \mathbb{R}$ that extends to a real character on G there exists a discrete character $\psi\colon G \to \mathbb{R}$ extending ϕ with kernel of type F_{n-1}.

The strategy of the proof consists in iteratively identifying a discrete character on G_{j+1} with kernel of increasing regularity, and induct on $0 \leq j \leq n-1$.

The outcome of Theorems 4.1 and 4.2 is that we can, under the assumption that the sequences in Equation (3) have excessive homology, identify algebraic fibrations with (higher) finiteness property: for instance, a poly-free group of length 3 with required excessive homology properties would contain an algebraic fiber of type F_2, namely finitely presented.

One may want to ask if this result is optimal. The quick answer is negative: for instance, $G = F_2 \times F_2 \times \mathbb{Z}$ admits an obvious fibration whose fiber has type F. Note that, in this case, the Euler characteristic of the group G is zero. In fact, at least for poly-free and poly-surface groups the nonvanishing of the Euler characteristic is the only condition required for optimality of the results above.

This is a consequence of the following result, that originates from [**Lu02**, Theorem 7.2(5)], and which is further discussed in [**Fi21**, **LIMP21**].

THEOREM 4.3 ([**Lu02**]). *Let G be a f.g. group and assume that the ith L^2-Betti number $b_i^{(2)}(G)$ is nonzero: then the fibers of all algebraic fibrations of G cannot have type $FP_i(\mathbb{Q})$.*

This result has relevence in the case of poly-free and poly-surface groups in light of the following known fact, whose proof can be found, e.g., in [**GN21**, **KV23**].

PROPOSITION 4.1. *Let G be a poly-free (respectively, poly-surface) group of length n. Then $b_i^{(2)}(G) = 0$ for $0 \leq i \leq n-1$, and*

$$b_n^{(2)}(G) = (-1)^n \chi(G) = (-1)^n \prod_{i=1}^{n} \chi(\Gamma_i).$$

(Note that for poly-surface groups this statement amounts to the fact that an aspherical manifold carrying G as fundamental group satisfies the Singer Conjecture.)

It follows that, unless one of the factor groups is $\mathbb{Z}$ (in the poly-free case) or $\mathbb{Z}^2$ (in the surface case) then an algebraic fiber cannot have type $FP_n(\mathbb{Q})$ nor, *a fortiori*, F_n. In particular, the fundamental group G of a surface bundle over a surface where both base and fiber have genus at least 2 and the first Betti number of the total space is strictly bigger than that of the base admits an algebraic fibration, so that $\Sigma^1(G) \neq \emptyset$, consistently with the case of Kodaira fibrations certified by Theorem 3.1.

There is an interesting class of groups where the assumption on excessive homology is easily verified, namely the pure braid groups. The pure mapping class group $P(n+1)$ in $n \geq 1$ strand is defined as the kernel of the epimorphism $B(n+1) \to S_{n+1}$ from the braid group in $n+1$ strands to the symmetric group, and geometrically corresponds to braids that do not permute their ends. While the BNSR invariants of the braid group $B(n+1)$ are rather uninteresting ($H_1(B(n+1);\mathbb{Z}) = \mathbb{Z}$

and it is not hard to show that the commutator subgroup has type F_∞, hence $\Sigma^m(B(n+1)) = S(B(n+1))$ for all m) they are much richer for $P(n+1)$. For these groups, the first BNSR invariant has been completely elucidated in [**KMM15**], and some partial results on higher BNSR invariants appear in [**Za17**] using Morse theory. We can essentially recast analog results using Theorems 4.1 and 4.3.

THEOREM 4.4 ([**Za17, KV23**]). *Let $G = P(n+1)$ be the pure braid group on $n+1$ strands; then for all $1 \leq m \leq n-2$, G contains an algebraic fibration with fiber of type F_m but not of type FP_{m+1}.*

The key points of our proof are the following. First, interpreting $P(n+1)$ as the fundamental group of the ordered configuration space of $(n+1)$ points on the plane, we have as a consequence of the Fadell–Neuwirth fibration (a point-forgetting map) a way to describe $P(n+1)$ iteratively as

$$1 \longrightarrow F_q \longrightarrow P(q+1) \longrightarrow P(q) \longrightarrow 1, \tag{4}$$

which, together with the fact that $P(2) = \mathbb{Z}$, allows one to interpret $P(n+1)$ as poly-free group of length n, with normal filtration given by

$$G_i = \operatorname{Ker}(P(n+1) \to P(n+1-i))$$

and factor groups $\Gamma_i = F_{n+1-i}$. Note that this entails that $\chi(P(n)) = 0$. However, we have an isomorphism

$$P(n+1) \cong K(n+1) \times \mathbb{Z},$$

where $K(n+1) \cong P(n+1)/Z(P(n+1))$ is isomorphic to the penultimate term $G_{n-1} = \operatorname{Ker}(P(n+1) \to P(2))$ of the filtration of $P(n+1)$, see [**FM12**, Section I.9]. The group $K(n+1)$ is then poly-free of length $n-1$, and none of its factor groups is free abelian, so its $(n-1)$th L^2-Betti number does not vanish. Finally, it is well known that the monodromy of all the extensions in Equation (4) act trivially on the homology of the fiber, so homologically $P(n+1)$ (and $K(n+1)$) behaves as a direct product as far as the first homology group is concerned. It is a straightforward consequence of this fact that the excessive homology in the sequences in Equation (3) is nontrivial (actually, maximal) so we can apply Theorem 4.1 to $K(n+1)$ (hence to $P(n+1)$) and obtain the result desired.

Applications of the machinery of Theorems 4.2 and 4.3 allows one also to get some information, via purely group-theoretical techniques, on higher BNSR invariants of some Kähler groups. We state an instance of this.

THEOREM 4.5 ([**KV23, LIP23**]). *There exist (complex projective) iterated Kodaira fibrations of complex dimension n with (Kähler) fundamental group G whose BNSR invariants are properly nested in the form*

$$\Sigma^0(G) \supsetneq \Sigma^1(G) \supsetneq \ldots \supsetneq \Sigma^{n-1}(G) \supsetneq \Sigma^n(G).$$

(Much as in the case of the pure braid groups, all the "jumps" can be obtained by using algebraic fibrations.) A proof of this theorem using the results of this paper appears in [**KV23**], while a proof based on a different approach appears in [**LIP23**].

It is interesting to compare at this point the statements of Theorems 4.1 and 4.2 to those obtained by Kielak and Fisher in the realm of virtually RFRS group. Recall the following definition.

DEFINITION 4.2. A group H is residually finite rationally solvable (RFRS) if there exists a normal filtration $H_i \trianglelefteq_f H$ such that $\cap_i H_i = \{1\}$ and the maps $H_i \to H_i/H_{i+1}$ factorize through the maximal free abelian quotient maps.

In this realm Kielak ([**Kie20**] for $n = 1$) and then Fisher ([**Fi21**] for $n > 1$) proved the following.

THEOREM 4.6 ([**Kie20, Fi21**]). *Let H be a virtually RFRS group of type $FP_n(\mathbb{Q})$. Then there exist a finite index subgroup of H which admits an algebraic fibration with fiber of type $FP_n(\mathbb{Q})$ if and only if the L^2-Betti numbers satisfy $b_i(H) = 0$ for $0 \leq i \leq n$.*

The obstruction to higher regularity, of course, is consistent with what is prescribed by Theorem 4.3. On the other hand, one may ask about the overlap between the statements of Theorems 4.1 or 4.2 and that of Theorem 4.6 as far as the construction of an algebraic fibration with the regularity prescribed. While a complete picture still eludes us, the overlap in the domain of application of these two types of results is likely to be quite limited. For instance, while direct products of virtually RFRS groups (e.g., free groups) are virtually RFRS, it is possible to prove (for rather nontrivial reasons; see [**KVW21**]) that there are free-by-free groups that are not virtually fibered, hence not virtually RFRS. This lends to the expectation that poly-free groups (and, more generally, nontrivial extensions) likely fail to be virtually RFRS.

One way to use the results discussed here is in the realm of higher coherence of groups. Recall that a group G is called *coherent* if all its finitely generated subgroups are finitely presented. We say that a group is m-coherent if all its subgroups of type F_m are of type F_{m+1}. A combination of Theorem 4.2 and Theorem 4.3 yields the following corollary.

COROLLARY 4.1. *Let G be a poly-free or a poly-surface of length n satisfying the excessive homology conditions of Equation* (3) *with nonvanishing Euler characteristic: then for all $0 \leq m \leq n-1$, G is not m-coherent.*

By definition, m-coherence descends to subgroups, so to prove that a group is not m-coherent we just need to find a poison subgroup that fails this property. Corollary 4.1 is, in essence, a recipe to build poison groups out of (suitable) poly-free and poly-surface groups without abelian factor groups. This idea, of course, is not new, and the use of embeddings of $F_2 \times F_2$ (or other direct products) is a simple but useful tool to prove incoherence.

Unsurprisingly, however, direct products of free (or surface) groups cannot always be used to get optimal results, so the strategy outlined above adds some new results that are out of reach otherwise. An instance is the following application to automorphism groups of free groups.

THEOREM 4.7 ([**V23**]). *For $n \geq 3$ there exists an algebraically fibered poly-free group of nonvanishing Euler characteristic of length $2n-2$ (respectively, $2n-3$), which embeds in* $\mathrm{Aut}(F_n)$ *(respectively,* $\mathrm{Out}(F_n)$*), such that the sequences in Equation* (3) *have excessive homology. Therefore,* $\mathrm{Aut}(F_n)$ *(respectively,* $\mathrm{Out}(F_n)$*) is not m-coherent for $0 \leq m \leq (2n-3)$ (respectively, $0 \leq m \leq (2n-4)$).*

Note that direct products of free groups of "high enough" length do not embed by [**HW20**]. Details of the proof, which hinge on the embedding of suitable (non product) poly-free groups as in the statement, due to [**BKK02**], can be found in [**V23**].

We finish this discussion on coherence with an observation: it has been conjectured (see [**Wi19**]) that the assumption on excessive homology for free-by-free groups is not necessary, even when iterated over finite index subgroups. Stated otherwise, for these groups, there should be witnesses to incoherence that are not necessarily (not even virtually) algebraic fibers. It is probably not too far fetched to conjecture the following generalization of Corollary 4.1, whose proof could prove rather challenging.

CONJECTURE 4.1. *Let G be a poly-free or a poly-surface of length n with non-vanishing Euler characteristic: then for all $0 \leq m \leq n-1$, G is not m-coherent.*

5. Profinite groups

5.1. Pro-$\mathcal{C}$ groups. Our main reference for profinite groups is the Ribes and Zalesskii book [**RZ10**]. In what follows we will generalize some of the results previously discussed to pro-$\mathcal{C}$ groups, where $\mathcal{C}$ is a class of finite groups. We briefly recall the definition.

DEFINITION 5.1. A pro-$\mathcal{C}$ group G is the inverse limit of its finite quotients that are in the class $\mathcal{C}$, i.e.,

$$G = \varprojlim G/U,$$

where $G/U \in \mathcal{C}$. If $\mathcal{C}$ is the class of all finite groups we say the group is profinite. If $\mathcal{C}$ is the class of all finite p-groups, were p a natural prime number, we say the group is pro-p.

A pro-$\mathcal{C}$ group is a topological group, where the above subgroups U form a nested set of open neighborhoods of 1_G. Alternatively, G embeds as a closed subgroup in the direct product $\prod G/U$, where each G/U is a topological group with the discrete topology and the topology of $\prod G/U$ is the product topology. Every pro-$\mathcal{C}$ group is compact and totally disconnected.

By definition, a subset X of a pro-$\mathcal{C}$ group G is a (topological) generating set if there is no proper closed subgroup of G that contains X. For an abstract group we have that every element of the group is a finite word on the set of generators and their inverses but this breaks completely in the category of profinite groups. But still as in the case of abstract groups the nice pro-$\mathcal{C}$ groups are the (topologically) finitely generated ones.

We note that if U is an open subgroup of a pro-$\mathcal{C}$ group G then we have a decomposition as a disjoint union $G = \cup_{t \in T} Ut$, where T is transversal. Then since each Ut is an open subset of G and G is compact, we have that T is finite, i.e., U has finite index in G. The converse holds for (topologically) finitely generated profinite groups, i.e., every subgroup of finite index in a (topologically) finitely generated profinite group is open. This is a deep result by Nikolov and Segal (see [**NS2-07**] and [**NS2-07**]) and it means that the topology of every (topologically) finitely generated profinite group is uniquely determined by its underlying algebraic structure.

From now on we restrict to pro-p groups G. They form a special class of profinite groups and are much better behaved compared to the general class of profinite groups. A pro-p group is procyclic if it is (topologically) generated by one element. The only infinite procyclic pro-p group is the group of the p-adic numbers $\mathbb{Z}_p$ that is the inverse limit $\varprojlim \mathbb{Z}/p^i\mathbb{Z}$.

We discuss briefly the Frattini subgroup of a pro-p group; for more details the reader can check in [**RZ10**, Section 2.8]. By definition the Frattini subgroup $\Phi(G)$ of a pro-p G is the intersection of all maximal closed subgroups of G. The Frattini subgroup $\Phi(G)$ is the pro-p subgroup of G (topologically) generated by the set

$$\{g^p \mid g \in G\} \cup \{[g_1, g_2] = g_1^{-1} g_2^{-1} g_1 g_2 \mid g_1, g_2 \in G\}.$$

It turns out that a subset X of G is a (topological) generating set of G if and only if the image of X in $G/\Phi(G)$ is a (topological) generating set. In particular, G is (topologically) finitely generated if and only if $G/\Phi(G)$ is finite, and in this case the minimal number of generators of G is the dimension of $G/\Phi(G)$ as a vector space over the field with p elements.

5.2. Finitely presented pro-p groups. For an abstract group H the pro-p completion $\widehat{H}_p$ of H is the inverse limit of all p-finite quotients of H. There is a natural map

$$H \longrightarrow \widehat{H}_p$$

that is injective precisely when H is residually p-finite.

Let F_0 be a free abstract group with free basis X. We assume that X is finite. The pro-p completion $F(X)$ of F_0 is the free pro-p group with a basis X. It satisfies the universal property that defines abstract free groups but now in the category of pro-p groups, i.e., if G is a pro-p group and $\alpha\colon X \to G$ is a set theoretic map, then there is a unique homomorphism of pro-p groups

$$f_\alpha : F(X) \longrightarrow G$$

that extends α. Recall that a homomorphism of pro-p groups by definition is a continuous group homomorphism.

For example, $\mathbb{Z}_p$ is a free pro-p group with basis with one element, as it is the pro-p completion of $\mathbb{Z}$. More generally, there is a definition of free pro-$\mathcal{C}$ group with an infinite basis X, where X is a profinite space, i.e., an inverse limit of finite sets; see [**RZ10**, Chapter 3].

Let G be a (topologically) finitely generated pro-p group. Then G is a pro-p quotient of the free pro-p group $F(X)$, with some finite X, i.e., there is an epimorphism of pro-p groups

$$\pi\colon F(X) \to G.$$

We say that G is finitely presented (as a pro-p group or in the category of pro-p groups) if there is a finite subset $R \subseteq Ker(\pi)$ such that $Ker(\pi)$ is the smallest closed normal subgroup of $F(X)$ that contains R. We write

$$G = \langle X \mid R \rangle_p.$$

If H is an abstract group that has finite presentation in the category of abstract groups $\langle X \mid R \rangle$, then

$$\widehat{H}_p = \langle X \mid R \rangle_p,$$

in particular, the pro-p completion of a finitely presented abstract group is finitely presented (in the pro-p sense). But if the pro-p completion of a finitely generated abstract group H_0 is finitely presented (as a pro-p group), this does not imply that H_0 is finitely presented as an abstract group, see [**Ki2-99**, Proposition I] for $p > 2$, H_0 finitely generated metabelian.

5.3. Homology, cohomology, and finiteness properties of pro-p groups. Let G be a pro-p group. By definition, the completed group algebra with coefficients in $\mathbb{Z}_p$ is

$$\mathbb{Z}_p[[G]] = \varprojlim \frac{\mathbb{Z}}{p^i\mathbb{Z}}[G/U],$$

where the inverse limit is over all $i \geq 1$ and U open subgroups of G. Similarly, we have

$$\mathbb{F}_p[[G]] = \mathbb{Z}_p[[G]]/p\mathbb{Z}_p[[G]] = \varprojlim \mathbb{F}_p[[G/U]]$$

where $\mathbb{F}_p$ is the field with p elements and the inverse limit is again over all open subgroups U of G. The completed group algebra plays in the category of pro-p groups the same role as the ordinary group algebra in the category of abstract groups.

The completed group algebras $\mathbb{Z}_p[[G]]$ and $\mathbb{F}_p[[G]]$ are pro-p rings, so they are topological rings. But they have one extremely nice property, they are *local* rings, i.e., they have a unique maximal ideal, that is, the kernel of the obvious augmentation map to $\mathbb{F}_p$ that maps G to 1. Because of this there is a pro-p version of the Nakayama lemma: if V is a pro-p $R[[G]]$-module, where $R = \mathbb{Z}_p$ or $\mathbb{F}_p$, then V is finitely generated if and only if $V/V\Omega$ is finite, where Ω is the unique maximal ideal of $R[[G]]$. Note that such a property does not hold in general for modules over abstract groups.

In the category of pro-p modules over pro-p rings we can develop homology and cohomology theory as in the abstract case; see [**RZ10**]. There are notions of free modules, injective modules, free resolutions, and projective resolutions. This allows us to define homology groups $H_i(G,V)$ and cohomology groups $H^i(G,W)$, where V is a pro-p $\mathbb{Z}_p[[G]]$-module and W is a discrete $\mathbb{Z}_p[[G]]$-module.

By [**RZ10**, Theorem 7.7.4] for a pro-p group G the following conditions are equivalent:

(a) the cohomological p-dimension $cd_p(G) \leq 1$;
(b) $H^2(G,\mathbb{F}_p) = 0$;
(c) G is a free pro-p group;
(d) G is a projective group.

As a corollary it follows that every closed subgroup of a free pro-p group G is free pro-p.

Note that for every pro-p group G we have

$$H_1(G,\mathbb{F}_p) \simeq G/\Phi(G).$$

Thus G is (topologically) finitely generated if and only if $dim_{\mathbb{F}_p} H_1(G,\mathbb{F}_p) < \infty$. This is equivalent to $\dim_{\mathbb{F}_p} H^1(G,\mathbb{F}_p) < \infty$.

If G is (topologically) finitely generated, then it is finitely presented (in pro-p sense) if and only if $dim_{\mathbb{F}_p} H_2(G,\mathbb{F}_p) < \infty$. This is equivalent to $dim_{\mathbb{F}_p} H^2(G,\mathbb{F}_p) < \infty$.

It is interesting to note that geometric methods, which so often work nicely for abstract groups, normally do not apply for pro-p groups. But homological and cohomological methods work nicely.

Next, we proceed with the definition of finiteness properties, which parallels that of Definition 2.1.

DEFINITION 5.2. Let G be a pro-p group and let n be a nonnegative integer. We say that G has type FP_n if $\mathbb{Z}_p$, as a trivial $\mathbb{Z}_p[[G]]$-module, admits a projective resolution by projective pro-p $\mathbb{Z}_p[[G]]$-modules with all morphism being continuous,

$$\ldots \longrightarrow \widetilde{P}_i \longrightarrow \widetilde{P}_{i-1} \ldots \longrightarrow \widetilde{P}_0 \longrightarrow \mathbb{Z}_p \longrightarrow 0,$$

with the first n terms finitely generated as pro-p $\mathbb{Z}_p[[G]]$-module.

By the fact that $\mathbb{Z}_p[[G]]$ is a local ring it follows that the following conditions are equivalent:

(a) G is of type FP_n
(b) all pro-p homology groups $H_i(G, \mathbb{F}_p)$ are finite for $i \leq n$;
(c) all pro-p cohomology groups $H^i(G, \mathbb{F}_p)$ are finite for $i \leq n$.

Indeed, the equivalence of the first two conditions is the King criterion Theorem 6.2 and the equivalence of the last two conditions follows by the Pontryagin duality; see [**RZ10**, Proposition 6.3.6]. Thus G is finitely presented (as a pro-p group) if and only if it is of type FP_2. Thus there is no pro-p version of the Bestvina–Brady example mentioned before. So this is another example that pro-p groups behave differently from abstract groups.

5.4. Golod–Shafarevich inequality for pro-p groups. For a pro-p group define $d(G)$ the minimal number of (topological) generators. In [**W91**] Wilson proved the following pro-p version of the Golod–Shafarevich inequality.

THEOREM 5.1 ([**W91**]). *Let G be a pro-p group with a finite pro-p presentation with n generators and r relations and $d = d(G) > 1$. Then either*

$$r \geq n + \frac{1}{4}d^2 - d$$

or for each finitely generated discrete dense subgroup H of G there is a closed normal subgroup K of G such that HK/K is an infinite, torsion group.

The following result was proved in [**W91**] as a corollary of Theorem 5.1.

THEOREM 5.2 ([**W91**]). *Let G be a finitely presented pro-p group and assume that there does not exist a finitely generated discrete subgroup H of G normalising K and a closed subgroup K of G such that HK/K is an infinite, torsion group. Then*

(a) *there is a constant k such that for each open subgroup U of G we have $d(U) \leq k[G : U]^{\frac{1}{2}}$;*
(b) *if N is any closed normal subgroup of G such that $G/N \simeq \mathbb{Z}_p$ then N is (topologically) finitely generated.*

The assumptions of Theorem 5.2 apply for soluble groups since every finitely generated, torsion, soluble group is finite. Part (b) from Theorem 5.2 is very striking. Such a result definitely does not hold for abstract groups, as there are many examples of splittings of a discrete group as a descending or ascending HNN extension, where the associated subgroups do not coincide. There is a pro-p version of the HNN construction but in the category of pro-p groups we do not have ascending or descending pro-p HNN extensions with no coinciding associated subgroups, since this cannot be done for finite p-groups.

One corollary of Theorem 5.2 is that for a finitely presented metabelian pro-p group G we have that if N is any closed normal subgroup of G such that $G/N \simeq \mathbb{Z}_p$ then N is (topologically) finitely generated. But this property does not characterise finite presentability. Finite presentability was classified later by King in [**Ki2-99**] and will be discussed in the next section.

6. Metabelian groups

In this section we compare the theory of finitely presented metabelian abstract and pro-p groups.

6.1. Finitely presented abstract, metabelian groups. Suppose H is a finitely generated abstract metabelian group. Then there is a short exact sequence of abstract groups

$$1 \longrightarrow A_0 \longrightarrow H \longrightarrow Q_0 \longrightarrow 1$$

with A_0 and Q_0 abelian. Here we view A_0 as a $\mathbb{Z}Q_0$-module via the Q_0-action induced by conjugation and the additive structure of A_0 is the underlying group operation restricted to A_0.

In [**BS80**] Bieri and Strebel define an invariant $\Sigma_{A_0}(Q_0)$ that is a subset of the character sphere

$$S(Q_0) = Hom(Q_0, \mathbb{R}) \setminus \{0\}/\sim,$$

where $\chi_1 \sim \chi_2$ if there is a positive real number r such that $\chi_1 = r\chi_2$. A_0 is said to be 2-tame if

$$S(Q_0) = \Sigma_{A_0}(Q_0) \cup -\Sigma_{A_0}(Q_0).$$

THEOREM 6.1 ([**BS80**]). *The following conditions are equivalent:*

(a) *H is finitely presented;*
(b) *H is of type FP_2;*
(c) *A_0 is 2-tame as $\mathbb{Z}Q_0$-module.*

Later on, the above invariant was generalized by Bieri, Neumann, and Strebel to the invariant $\Sigma^1(G)$ described in Section 2. Condition (c) from the above theorem can be restated as $S(H) = \Sigma^1(H) \cup -\Sigma^1(H)$.

The fact that (b) implies (c) was proved by a very beautiful geometric argument that uses algebraic topology, in particular covering maps and van Kampen's theorem. The fact that (c) implies (a) is an algebraic calculation, a very nontrivial one. Note that one unexpected corollary of the above result is that finite presentability does not depend on the type of the extension, i.e., does not depend on the element of $H^2(Q_0, A_0)$ that describes the group H but only on the structure of A_0 as a $\mathbb{Z}Q_0$-module. This is somewhat surprising since in general, i.e., outside the class of metabelian groups, we expect that the finite presentability depends on the type of the extension.

There is a curious link between $\Sigma_{A_0}(Q_0)$ and valuation theory from commutative algebra. This was used in [**BGr84**] to show that $S(Q_0) \setminus \Sigma_{A_0}(Q_0)$ is a rationally defined spherical polyhedron. This result itself turned out important and quite useful in the modern area of tropical geometry.

6.2. Finitely presented metabelian pro-p groups. For general finitely presented groups G in [**Ki1-99**], King proved the following criterion.

THEOREM 6.2 ([**Ki1-99**]). *Suppose that G is a pro-p group, R is either $\mathbb{F}_p$ or $\mathbb{Z}_p$, and N a normal pro-p subgroup of G such that $R[[G/N]]$ is left and right Noetherian. Then G is of type FP_m as a pro-p group if and only if the homology groups $H_i(N, R)$ are finitely generated as pro-p $R[[G/N]]$-modules for all $i \leq m$, where the G/N action is induced by the conjugation action of G on N.*

As stated before geometric methods do not pass to pro-p groups. Still there is a complete classification of finitely presented metabelian pro-p groups G. As a corollary of the above theorem King obtained the following classification

THEOREM 6.3. *Let $1 \to A \to G \to Q \to 1$ be a short exact sequence of pro-p groups, where A and Q are abelian and G is (topologically) finitely generated. Then the following conditions are equivalent:*

(a) *G is finitely presented (as a pro-p group)*
(b) *$H_2(A, \mathbb{Z}_p) \simeq A\widehat{\wedge}A$ is finitely generated as a $\mathbb{Z}_p[[Q]]$-module via the diagonal Q-action.*

We note that in the above theorem the fact that G is (topologically) finitely generated and Q is (topologically) finitely generated abelian, hence (topologically) finitely presented imply that A is (topologically) finitely generated as a $\mathbb{Z}_p[[Q]]$-module.

In the above theorem $A\widehat{\wedge}A$ is the exterior product in the category of pro-p modules. It is the inverse limit of $(A/U) \wedge (A/U)$ over all pro-p $\mathbb{Z}_p[[Q]]$-submodules U of A with A/U p-finite. Note that Theorem 6.3 is a direct corollary of Theorem 6.2 applied for $N = A$. In this case $Q \simeq G/A$ is a (topologically) finitely generated abelian pro-p group, hence both rings $\mathbb{Z}_p[[Q]]$ and $\mathbb{F}_p[[Q]]$ are Noetherian.

Note that as observed before a pro-p $\mathbb{Z}_p[[Q]]$-module V is (topologically) finitely generated if and only if $V/V\Omega$ is finite where Ω is the unique maximal ideal of $\mathbb{Z}_p[[Q]]$. In particular, condition (b) is equivalent to $A\widehat{\wedge}A/(A\widehat{\wedge}A)\Omega$ is finite.

It is interesting to note that in [**Ki2-99**] King defines an invariant that plays in the theory of pro-p metabelian groups the same role as the complement $\Sigma^c_{A_0}(Q_0) = S(Q_0) \setminus \Sigma_{A_0}(Q_0)$ of the first Bieri–Strebel invariant $\Sigma_{A_0}(Q_0)$. He calls the invariant geometric in the title of the paper but actually this invariant is everything but geometric. In some sense King's invariant resembles the link between $\Sigma^c_{A_0}(Q_0)$ and valuation theory, as suggested by Bieri and Groves in [**BGr84**], but all this was adapted in the context of pro-p groups.

One of the corollaries of the main results from [**BS80**] is that if H is a finitely presented abstract group without free noncyclic subgroups then every metabelian quotient of H is finitely presented. The main part of the proof uses the topological methods mentioned before and thus is not applicable to pro-p groups. The same statement makes sense for pro-p groups but it is still an open problem, i.e., it is open whether G is a finitely presented pro-p group without free nonprocyclic pro-p subgroups then the maximal metabelian pro-p quotient of G is finitely presented pro-p.

6.3. The King invariant. Let Q be a finitely generated, abelian, pro-p group. We denote by $\overline{\mathbb{F}}_p$ the algebraic closure of $\mathbb{F}_p$ and by $\overline{\mathbb{F}}_p[[t]]^\times$ the multiplicative group of all invertible elements in the power series ring $\overline{\mathbb{F}}_p[[t]]$. Following King in [**Ki2-99**] we set

$$T(Q) = \{\chi : Q \longrightarrow \overline{\mathbb{F}}_p[[t]]^\times \mid \chi \text{ is a continuous homomorphism}\},$$

here $\overline{\mathbb{F}}_p[[t]]^\times$ is considered as a topological group with topology induced by the sequence of ideals $(t) \supseteq (t^2) \supseteq \ldots \supseteq (t^i) \supseteq \ldots$. For $\chi \in T(Q)$ we define by

$$\overline{\chi} : \mathbb{Z}_p[[Q]] \longrightarrow \overline{\mathbb{F}}_p[[t]]$$

the unique continuous ring homomorphism that extends the continuous homomorphism χ.

Let A be a finitely generated pro-p $\mathbb{Z}_p[[Q]]$-module. In [**Ki2-99**] King defined the following invariant

$$\Delta(A) = \{\chi \in T(Q) \mid ann_{\mathbb{Z}_p[[Q]]}(A) \subseteq Ker(\overline{\chi})\}.$$

The following result is an alternative classification of the finitely presented metabelian pro-p groups given by King in [**Ki2-99**].

THEOREM 6.4 ([**Ki2-99**]). *Let $1 \to A \to G \to Q \to 1$ be a short exact sequence of pro-p groups, where G is a (topologically) finitely generated pro-p group and A and Q are abelian. Then G is a finitely presented pro-p group if and only if $\Delta(A) \cap \Delta(A)^{-1} = \{1\}$.*

7. Algebraic fibering for pro-p groups

The main purpose of this section is to discuss the pro-p versions of some of the results in Section 4. The proofs can be found in [**Ko22**].

THEOREM 7.1 ([**Ko22**]). *Let $1 \to K \to G \to \Gamma \to 1$ be a short exact sequence of pro-p groups and $n_0 \geq 1$ be an integer such that*

(1) *G and K are of type FP_{n_0},*
(2) *Γ^{ab} is infinite,*
(3) *there is a normal pro-p subgroup N of K such that $G' \cap K \subseteq N$, $K/N \simeq \mathbb{Z}_p$ and N is of type FP_{n_0-1}.*

Then there is a normal pro-p subgroup M of G such that $G/M \simeq \mathbb{Z}_p$, $M \cap K = N$ and M is of type FP_{n_0}. Furthermore if K, G and N are of type FP_∞ then M can be chosen of type FP_∞.

We note that as in the case of abstract groups, in the pro-p case condition (1) from Theorem 7.1 implies that Γ is of type FP_{n_0}. Actually, it suffices that G is of type FP_{n_0} and K is of type FP_{n_0-1}.

We say that a pro-p group G is coherent (as a pro-p group or in the category of pro-p groups) if every finitely generated pro-p subgroup of G is finitely presented as a pro-p group. This generalizes to higher dimensions: a pro-p group G is n-coherent if any pro-p subgroup of G that is of type FP_n is of type FP_{n+1}. In this notation a pro-p group is 1-coherent if and only if it is coherent (in the category of pro-p groups).

COROLLARY 7.1 ([**Ko22**]). *Let K, Γ and $G = K \rtimes \Gamma$ be pro-p groups and $n_0 \geq 1$ be an integer such that*

(1) *Γ is finitely generated free pro-p but not procyclic,*
(2) *K is of type FP_{n_0},*
(3) *there is a normal pro-p subgroup N of K such that $G' \cap K \subseteq N$, $K/N \simeq \mathbb{Z}_p$ and N is of type FP_{n_0-1} but is not of type FP_{n_0}.*

Then there is a normal pro-p subgroup M of G such that $G/M \simeq \mathbb{Z}_p$, $M \cap K = N$ and M is of type FP_{n_0} but is not of type FP_{n_0+1}. In particular, G is not n_0-coherent.

The proofs of the above results are not geometric as in the abstract case, since we do not have geometric tools here as Σ-theory. But we use homological methods that are more appropriate for the category of pro-p groups, including the Lyndon–Hochschild–Serre spectral sequence. We also reduce several problems to studying finite generation of some particular pro-p modules and in some cases use King's Δ-invariant. Though it has no geometric flavor it should be considered as the pro-p version of the complement of the Bieri–Neumann–Strebel Σ^1 invariant.

In [**Ko22**] we use the following result that is a pro-p version of a result of Kuchkuck from [**K14**]. It is interesting to note that the map θ from the statement of Lemma 7.1 is not supposed to be surjective or injective. Lemma 7.1 is a very surprising result that needs more attention and deserves consideration not only as an auxiliary result used in the proof of Theorem 7.1.

LEMMA 7.1 ([**Ko22**]). *Let $n \geq 1$ be a natural number,*

$$1 \longrightarrow A \longrightarrow B \longrightarrow C \longrightarrow 1$$

a short exact sequence of pro-p groups with A of type FP_n and C of type FP_{n+1}. Assume there is another short exact sequence of pro-p groups

$$1 \longrightarrow A \longrightarrow B_0 \longrightarrow C_0 \longrightarrow 1$$

with B_0 of type FP_{n+1} and that there is a homomorphism of pro-p groups $\theta : B_0 \to B$ such that $\theta|_A = id_A$, i.e., there is a commutative diagram of homomorphisms of pro-p groups

$$\begin{array}{ccccccccc} 1 & \longrightarrow & A & \longrightarrow & B_0 & \xrightarrow{\pi_0} & C_0 & \longrightarrow & 1 \\ & & \downarrow{\scriptstyle id_A} & & \downarrow{\scriptstyle \theta} & & \vdots{\scriptstyle \nu} & & \\ 1 & \longrightarrow & A & \longrightarrow & B & \xrightarrow{\pi} & C & \longrightarrow & 1. \end{array}$$

Then B is of type FP_{n+1}.

8. Coherence and incoherence for pro-p groups

In [**KW22**], Kropholler and Walsh asked and studied the question whether for F_k the free abstract group of rank k,

every group $H = F_n \rtimes F_k$ with $n, k \geq 2$ is not coherent.

As we mentioned, this question also arose in [**Wi19**]. This question and related topics were studied in [**KVW21**], [**FV23**]. The main point is to look for a subgroup H_0 of finite index in H, hence it is free-by-free, but with a better homological property, namely

$$H_0 = K_0 \rtimes \Gamma_0$$

with K_0 and Γ_0 free of finite rank, $d(\Gamma_0) > 1$ and the short exact sequence

$$1 \longrightarrow K_0 \longrightarrow H_0 \longrightarrow \Gamma_0 \longrightarrow 1 \text{ has excessive homology,}$$

which means that

$$rk(H^1(H_0, \mathbb{R})) > rk(H^1(\Gamma_0, \mathbb{R})).$$

Unfortunately, as shown in [**KVW21**], it is not always possible to find such H_0.

The situation for pro-p groups is quite different. It makes sense to consider a pro-p group

$$G = K \rtimes \Gamma,$$

where K and Γ are (topologically) finitely generated, free, nonprocyclic pro-p groups. Then Γ acts by conjugation on K and this gives an action of Γ on the abelianization of K, i.e., maximal abelian pro-p quotient of K, that is, $\mathbb{Z}_p^n$ for appropriate positive integer $n \geq 2$. This gives a continuous homomorphism

$$\rho\colon \Gamma \to GL_n(\mathbb{Z}_p).$$

Note that $GL_n(\mathbb{Z}_p)$ is a profinite group, that is, virtually pro-p. A pro-p group that (continuously) embeds in $GL_n(\mathbb{Z}_p)$ is pro-p analytic and cannot have nonprocyclic free pro-p subgroups (the class of analytic pro-p groups is an important class of pro-p groups; for more details, see [**DSMS**]). This is the big difference with the abstract case. The group $GL_n(\mathbb{Z})$ has many free noncyclic subgroups, but $GL_n(\mathbb{Z}_p)$ cannot have closed free nonprocyclic pro-p subgroups! In particular,

$$\rho \text{ is not injective.}$$

Then we can take Γ_0 a nonprocyclic, (topologically) finitely generated pro-p subgroup of $Ker(\rho)$ and consider the group

$$G_0 = K \rtimes \Gamma_0.$$

For this group we have that Γ_0 acts trivially on the abelianization of K. We write G_0' for the commutator subgroup of G_0, i.e., the closed subgroup (topologically) generated by $\{[g_1, g_2] \mid g_1, g_2 \in G_0\}$. Then

$$G_0' \cap K = K'$$

and using that for any normal pro-p subgroup N of K such that $K/N \simeq \mathbb{Z}_p$ we have that N is not (topologically) finitely generated, we can apply Corollary 7.1, i.e., we construct a pro-p subgroup of G_0 that is (topologically) finitely generated but not finitely presented (in pro-p sense). Thus G_0 is pro-p incoherent.

This argument can be generalized for a bigger class of groups denoted by $\mathcal{L}$. The class of pro-p groups $\mathcal{L}$ was defined in [**KZ11**] by Kochloukova and Zalesskii and should be considered as an attempt to generalize limit groups to the pro-p setting. It contains all (topologically) finitely generated free pro-p groups. A profinite version of it was studied by Zalesskii and Zapata in [**ZZ20**]. A pro-p group from $\mathcal{L}$ has many properties that resemble an abstract limit group. The class of abstract limit groups (over free groups) was studied independently by Sela and Kharlampovich and Myasnikov. It coincides with the class of all fully residually free abstract groups. Still, we do not have a good understanding what the theory of fully residually free pro-p groups should be. The definition of the class $\mathcal{L}$ was an attempt to use the extension of centralizers construction of Kharlampovich and Myasnikov to study pro-p groups.

COROLLARY 8.1 ([**Ko22**]). *Let* $1 \to K \to G \to \Gamma \to 1$ *be a short exact sequence of pro-p groups such that*

(1) K *is a nonabelian pro-p group from the class* $\mathcal{L}$*,*
(2) Γ *contains a nonabelian, free pro-p subgroup.*

Then G *is incoherent (in the category of pro-p groups). In particular, if* K *is a finitely generated free pro-p group with* $d(K) \geq 2$ *then* G *is incoherent (in the category of pro-p groups).*

We point out that not much is known for pro-p coherence. But in a recent paper Snopce and Zalesskii proved that some pro-p right angled Artin groups are coherent. More precisely, if Γ is a finite simplicial graph, the right angled Artin pro-p group G_Γ associated to Γ is defined by the pro-p presentation

$$\langle V(\Gamma) \mid g_1g_2 = g_2g_1 \text{when } g_1, g_2 \text{ are connected by an edge in } \Gamma\rangle_p.$$

This is the pro-p completion of the corresponding abstract right angled Artin group. The graph Γ is called chordal if every circuit of length at least 4 has a chord (i.e., there is an edge outside of the circuit that links two vertices from the circuit). Thus every circuit of minimal length should have precisely 3 edges.

PROPOSITION 8.1 ([**SZ22**]). *Let Γ be a chordal graph. Then every (topologically) finitely generated pro-p subgroup of G_Γ is of homological type FP_∞. In particular, G_Γ is coherent (in the category of pro-p groups).*

By [**S**] a Demushkin pro-p group G is a Poincaré duality group of dimension 2. Thus $H^i(G, \mathbb{F}_p)$ is finite for all i, $\dim H^2(G, \mathbb{F}_p) = 1$ and the cup product

$$\cup : H^i(G, \mathbb{F}_p) \times H^{2-i}(G, \mathbb{F}_p) \longrightarrow H^2(G, \mathbb{F}_p)$$

is a nondegenerated bilinear form for all $i \geq 0$.

There are two invariants associated to a Demushkin pro-p group: the minimal number of (topological) generators d and q that is either a power of the prime p or ∞. More on Demushkin pro-p groups can be found in [**W98**]. Demushkin pro-p groups were described in terms of presentations in [**De61**], [**De63**], [**La67**], and [**S71**]. We state below the classification of Demushkin pro-p groups due to Demushkin, Labute, and Serre.

THEOREM 8.1 ([**De61**], [**De63**]). *Let D be a Demushkin group with invariants d, q and suppose that $q \neq 2$. Then d is even and D is isomorphic to F/R, where F is a free pro-p group with basis $x_1, \dots, x_d$ and R is generated as a normal closed subgroup by*

$$x_1^q[x_1, x_2] \dots [x_{d-1}, x_d],$$

where for $q = \infty$ we define $x_1^\infty = 1$. Furthermore, all groups having such presentations are Demushkin.

THEOREM 8.2 ([**S71**]). *Let D be a Demushkin pro-2 group with invariants d, q and suppose that $q = 2$ and d is odd. Then D is isomorphic to F/R, where F is a free pro-2 group with basis $x_1, \dots, x_d$ and R is generated as a normal closed subgroup by*

$$x_1^2 x_2^{2^f}[x_2, x_3] \dots [x_{d-1}, x_d]$$

for some integer $f \geq 2$ or ∞. Furthermore, all groups having such presentations are Demushkin.

THEOREM 8.3 ([**La67**]). *Let D be a Demushkin pro-2 group with d even and $q = 2$. Then D is isomorphic to F/R, where F is a free pro-2 group with basis $x_1, \dots, x_d$ and R is generated as a normal closed subgroup either by*

$$x_1^{2^f+2}[x_1, x_2][x_3, x_4] \dots [x_{d-1}, x_d]$$

for some integer $f \geq 2$ or $f = \infty$, or by

$$x_1^2[x_1, x_2] x_3^{2^f}[x_3, x_4] \dots [x_{d-1}, x_d]$$

for some integer $f \geq 2$ or $f = \infty$, $d \geq 4$. Furthermore, all groups having such presentations are Demushkin.

As a corollary of Corollary 7.1 we obtain the following result on incoherence.

COROLLARY 8.2 ([**Ko22**]). *Let* $1 \to K \to G \to \Gamma \to 1$ *be a short exact sequence of pro-p groups such that*

(1) K *is a nonabelian pro-p RAAG or a nonsoluble Demushkin group,*
(2) Γ *is a nonabelian pro-p RAAG or a nonsoluble Demushkin group.*

Then G *is incoherent (in the category of pro-p groups).*

Let G be a finitely generated pro-p group. Define

$$Aut_0(G) = Ker(Aut(G) \longrightarrow Aut(G/\Phi(G))),$$

where $\Phi(G)$ is the Frattini subgroup of G. Then $Aut_0(G)$ is a pro-p subgroup of $Aut(G)$ of finite index. For a (topologically) finitely generated free pro-p group F the topological group $Aut(F)$ was studied by Lubotsky in [**L82**].

We are interested whether a pro-p version of the Gordon result from [**G04**] that the automorphism group of the free abstract group of rank 2 is incoherent holds.

COROLLARY 8.3 ([**Ko22**]). *Suppose that* K *is a finitely generated free pro-p group with* $d(K) \geq 2$. *If* $Out(K)$ *contains a pro-p free non-procyclic subgroup then* $Aut_0(K)$ *is incoherent (in the category of pro-p groups).*

As mentioned before by the results in [**BS80**] every metabelian quotient of a finitely presented abstract group that does not contain free noncyclic abstract subgroups is itself finitely presented. But we do not know whether a pro-p version of this holds. Using Corollary 8.3 and some results of Romankov from [**R92**] and [**R93**] the following result was proved in [**Ko22**].

COROLLARY 8.4 ([**Ko22**]). *Suppose that* K *is a finitely generated free pro-p group with* $d(K) \geq 2$. *Then either* $Aut_0(K)$ *is incoherent (in the category of pro-p groups) or the pro-p version of the Bieri–Strebel result does not hold.*

Recently, Marco Boggi has told us that the condition that $Out(K)$ contains a pro-p free nonprocyclic subgroup from Corollary 8.3 can be verified in many cases. We include an appendix that treats this condition.

Appendix: A lemma on the outer automorphism group of a free pro-p group

BY MARCO BOGGI

For a given group G, we denote by $\widehat{G}^{(p)}$ its pro-p completion and by $d(G)$ the cardinality of a minimal set of generators. We then have the following lemma.

LEMMA A.1. *Let* F *be a free group.*

(1) *If* $d(F) \geq 3$, *then* $\mathrm{Out}(\widehat{F}^{(p)})$ *contains a free pro-p group of rank* $d(F) - 1$.
(2) *If* $d(F) = 2$, *then* $\mathrm{Out}(\widehat{F}^{(2)})$ *contains a free pro-2 group of rank* 2.

PROOF.

(1) Let $S = S_{g,n}$ be a closed connected orientable surface of genus g from which n points have been removed and let $\Pi = \Pi_{g,n}$ be its fundamental group for a fixed base point $P \in S$. So that $\Pi_{0,n}$ is free of rank $n-1$ and $\Pi_{1,1}$ is free of rank 2. Let us assume that $2g - 2 + n > 0$ and that $n > 0$. Let $\Pi_P := \pi_1(S \setminus P, a)$, for some base point $a \in S \setminus P$. Note then that $d(\Pi_P) = d(\Pi) + 1$.

Let $\Gamma(S)$ be the mapping class group of the surface S and $\Gamma(S,P)$ be the mapping class group of the marked surface (S,P). There is a short exact sequence (the Birman sequence, cf. [**FM12**, Section 4.2]):

$$1 \longrightarrow \Pi \longrightarrow \Gamma(S,P) \longrightarrow \Gamma(S) \longrightarrow 1.$$

The representation $\rho_P\colon \Gamma(S,P) \to \mathrm{Aut}(\Pi)$ induced by the action (modulo isotopy) of elements of $\Gamma(S,P)$ on the marked surface (S,P) is faithful and is equivalent to the restriction of inner automorphisms from $\Gamma(S,P)$ to its normal subgroup Π. In particular, this representation identifies the inclusion $\Pi \subset \Gamma(S,P)$ in the Birman sequence with the inclusion $\mathrm{Inn}\,\Pi \subset \mathrm{Aut}(\Pi)$.

The representation $\rho_P^{\mathrm{out}}\colon \Gamma(S,P) \to \mathrm{Out}(\Pi_P)$, induced by the action (modulo isotopy) of elements of $\Gamma(S,P)$ on the surface $S \smallsetminus P$, is also faithful. This representation then identifies the subgroup Π of $\Gamma(S,P)$ with a subgroup of $\mathrm{Out}(\Pi_P)$.

This works also in the pro-p setting. The profinite topologies on $\mathrm{Aut}(\widehat{\Pi}^{(p)})$ and $\mathrm{Out}(\widehat{\Pi}_P^{(p)})$ induce profinite topologies on $\Gamma(S,P)$ via the representations ρ_P and ρ_P^{out} and the natural homomorphisms $\mathrm{Aut}(\Pi) \to \mathrm{Aut}(\widehat{\Pi}^{(p)})$ and $\mathrm{Out}(\Pi_P) \to \mathrm{Out}(\widehat{\Pi}_P^{(p)})$. In principle, these two topologies might be distinct but they are in fact equivalent (cf. [**Bo20**, Corollary 4.7]) and called the *p-congruence topology* on $\Gamma(S,P)$. The completion of $\Gamma(S,P)$ with respect to this topology is denoted by $\check{\Gamma}^{(p)}(S,P)$ and called the *p-congruence completion* of $\Gamma(S,P)$.

Since $\widehat{\Pi}^{(p)}$ is center free, there is a natural monomorphism $\widehat{\Pi}^{(p)} \cong \mathrm{Inn}(\widehat{\Pi}^{(p)}) \hookrightarrow \mathrm{Aut}(\widehat{\Pi}^{(p)})$. This implies that the restriction of the p-congruence topology on Π, via the embedding $\Pi \hookrightarrow \Gamma(S,P)$, is the full pro-$p$ topology. Hence, there is a natural monomorphism $\widehat{\Pi}^{(p)} \hookrightarrow \check{\Gamma}^{(p)}(S,P)$ and thus a natural monomorphism:

$$\widehat{\Pi}^{(p)} \hookrightarrow \mathrm{Out}(\widehat{\Pi}_P^{(p)}).$$

For $S = S_{0,n}$ and $n \geq 3$, this gives an embedding of a free pro-p group of rank $n-1$ in the outer automorphism group of a free pro-p group of rank n.

(2) By [**Bo24**, Propositions 2.5 and 4.1], the kernel of the natural representation $\Gamma(S_{1,1}) \cong SL_2(\mathbb{Z}) \to SL_2(\mathbb{Z}/2)$ is the direct product of a free group D_2 of rank 2, generated by the squares of Dehn twists (these correspond to the transvections of $SL_2(\mathbb{Z})$ via the above isomorphism), and the center of $\Gamma(S_{1,1})$, which is a cyclic group of order 2.

The conclusion then follows from the 2-congruence property for $\Gamma(S_{1,1})$ (cf. (i) [**Ho19**, Theorem A] or (ii) [**Bo20**, Theorem 1.4]). This in fact implies that the natural homomorphism $\widehat{D_2}^{(2)} \to \mathrm{Out}(\hat{\Pi}_P^{(2)})$, induced by the representation ρ_P^{out} : $\Gamma(S_{1,1}) \hookrightarrow \mathrm{Out}(\Pi_P)$, is also faithful.

Note that a similar argument would fail for primes $p \geq 11$, since the p-congruence subgroup property for $\Gamma(S_{1,1})$ does not hold in these cases by (ii) [**Ho19**, Theorem A]. □

Acknowledgments

It is a pleasure to thank Stefan Friedl, Rob Kropholler, and Genevieve Walsh for many conversations on the topics of this survey. The second author would also like to thank Luca Di Cerbo and Laurentiu Maxim for organizing a stimulating Session on the Singer–Hopf Conjecture in Geometry and Topology at the AMS Sectional Meeting of March 2023 in Atlanta, GA, in which many of the results contained in this survey were presented. We thank the referee for the numerous suggestions.

References

[BKK02] M. Bestvina, M. Kapovich, and B. Kleiner, *Van Kampen's embedding obstruction for discrete groups*, Invent. Math. **150** (2002), no. 2, 219–235, DOI 10.1007/s00222-002-0246-7. MR1933584

[BB97] M. Bestvina and N. Brady, *Morse theory and finiteness properties of groups*, Invent. Math. **129** (1997), no. 3, 445–470, DOI 10.1007/s002220050168. MR1465330

[BG10] R. Bieri and R. Geoghegan, *Sigma invariants of direct products of groups*, Groups Geom. Dyn. **4** (2010), no. 2, 251–261, DOI 10.4171/GGD/82. MR2595091

[BGr84] R. Bieri and J. R. J. Groves, *The geometry of the set of characters induced by valuations*, J. Reine Angew. Math. **347** (1984), 168–195. MR733052

[BNS87] R. Bieri, W. D. Neumann, and R. Strebel, *A geometric invariant of discrete groups*, Invent. Math. **90** (1987), no. 3, 451–477, DOI 10.1007/BF01389175. MR914846

[BS80] R. Bieri and R. Strebel, *Valuations and finitely presented metabelian groups*, Proc. London Math. Soc. (3) **41** (1980), no. 3, 439–464, DOI 10.1112/plms/s3-41.3.439. MR591649

[BR88] R. Bieri and B. Renz, *Valuations on free resolutions and higher geometric invariants of groups*, Comment. Math. Helv. **63** (1988), no. 3, 464–497, DOI 10.1007/BF02566775. MR960770

[Br94] K. S. Brown, *Cohomology of groups*, Graduate Texts in Mathematics, vol. 87, Springer-Verlag, New York, 1994. Corrected reprint of the 1982 original. MR1324339

[Br10] K. S. Brown, *Lectures on the cohomology of groups*, Cohomology of groups and algebraic K-theory, Adv. Lect. Math. (ALM), vol. 12, Int. Press, Somerville, MA, 2010, pp. 131–166. MR2655176

[Bo20] M. Boggi, *Congruence topologies on the mapping class group*, J. Algebra **546** (2020), 518–552, DOI 10.1016/j.jalgebra.2019.11.004. MR4035001

[Bo24] M. Boggi, *Notes on hyperelliptic mapping class groups*, Glasg. Math. J. **66** (2024), no. 1, 126–161, DOI 10.1017/s0017089523000381. MR4709062

[BU11] M. Burger, *Fundamental groups of Kähler manifolds and geometric group theory*, Astérisque **339** (2011), Exp. No. 1022, ix, 305–321. Séminaire Bourbaki. Vol. 2009/2010. Exposés 1012–1026. MR2906358

[Ca03] F. Catanese, *Fibred Kähler and quasi-projective groups*, Adv. Geom. **suppl.** (2003), S13–S27, DOI 10.1515/advg.2003.2003.s1.13. Special issue dedicated to Adriano Barlotti. MR2028385

[De10] T. Delzant, *L'invariant de Bieri-Neumann-Strebel des groupes fondamentaux des variétés kählériennes* (French, with French summary), Math. Ann. **348** (2010), no. 1, 119–125, DOI 10.1007/s00208-009-0468-8. MR2657436

[De61] S. P. Demuškin, *The group of a maximal p-extension of a local field* (Russian), Izv. Akad. Nauk SSSR Ser. Mat. **25** (1961), 329–346. MR123565

[De63] S. P. Demuškin, *On 2-extensions of a local field* (Russian), Sibirsk. Mat. Ž. **4** (1963), 951–955. MR161854

[DSMS] J. D. Dixon, M. P. F. du Sautoy, A. Mann, and D. Segal, *Analytic pro-p-groups*, London Mathematical Society Lecture Note Series, vol. 157, Cambridge University Press, Cambridge, 1991. MR1152800

[FM12] B. Farb and D. Margalit, *A primer on mapping class groups*, Princeton Mathematical Series, vol. 49, Princeton University Press, Princeton, NJ, 2012. MR2850125

[Fi21] S. Fisher, *Improved algebraic fibering*, preprint.

[FV23] S. Friedl and S. Vidussi, *BNS invariants and algebraic fibrations of group extensions*, J. Inst. Math. Jussieu **22** (2023), no. 2, 985–999, DOI 10.1017/S1474748021000438. MR4557911

[GN21] D. Gaboriau and C. Noûs, *On the top-dimensional ℓ^2-Betti numbers* (English, with English and French summaries), Ann. Fac. Sci. Toulouse Math. (6) **30** (2021), no. 5, 1121–1137, DOI 10.5802/afst.1695. MR4401387

[HW20] C. Horbez and R. D. Wade, *Commensurations of subgroups of* $\mathrm{Out}(F_N)$, Trans. Amer. Math. Soc. **373** (2020), no. 4, 2699–2742, DOI 10.1090/tran/7991. MR4069231

[Ho19] Y. Hoshi and Y. Iijima, *A pro-l version of the congruence subgroup problem for mapping class groups of genus one*, J. Algebra **520** (2019), 1–31, DOI 10.1016/j.jalgebra.2018.10.036. MR3880862

[G04] C. McA. Gordon, *Artin groups, 3-manifolds and coherence*, Bol. Soc. Mat. Mexicana (3) **10** (2004), no. Special Issue, 193–198. MR2199348

[Kie20] D. Kielak, *Residually finite rationally solvable groups and virtual fibring*, J. Amer. Math. Soc. **33** (2020), no. 2, 451–486, DOI 10.1090/jams/936. MR4073866

[Ki1-99] J. D. King, *Homological finiteness conditions for pro-p groups*, Comm. Algebra **27** (1999), no. 10, 4969–4991, DOI 10.1080/00927879908826743. MR1709190

[Ki2-99] J. D. King, *A geometric invariant for metabelian pro-p groups*, J. London Math. Soc. (2) **60** (1999), no. 1, 83–94, DOI 10.1112/S0024610799007693. MR1721817

[Ko22] D. H. Kochloukova, *Higher dimensional algebraic fiberings for pro-p groups*, Canad. J. Math., (2024), to appear.

[K-L18] D. H. Kochloukova and F. F. Lima, *Homological finiteness properties of fibre products*, Q. J. Math. **69** (2018), no. 3, 835–854, DOI 10.1093/qmath/hax063. MR3859210

[KZ11] D. H. Kochloukova and P. A. Zalesskii, *On pro-p analogues of limit groups via extensions of centralizers*, Math. Z. **267** (2011), no. 1-2, 109–128, DOI 10.1007/s00209-009-0611-y. MR2772243

[KMM15] N. Koban, J. McCammond, and J. Meier, *The BNS-invariant for the pure braid groups*, Groups Geom. Dyn. **9** (2015), no. 3, 665–682, DOI 10.4171/GGD/323. MR3420539

[KV23] D. H. Kochloukova and S. Vidussi, *Higher dimensional algebraic fiberings of group extensions*, J. Lond. Math. Soc. (2) **108** (2023), no. 3, 978–1003, DOI 10.1112/jlms.12773. MR4639944

[KVW21] R. Kropholler, S. Vidussi, and G. S. Walsh, *Virtual algebraic fibrations of surface-by-surface groups and orbits of the mapping class group*, J. Algebra **651** (2024), 1–18, DOI 10.1016/j.jalgebra.2024.03.030. MR4735761

[KW22] R. P. Kropholler and G. S. Walsh, *Incoherence and fibering of many free-by-free groups* (English, with English and French summaries), Ann. Inst. Fourier (Grenoble) **72** (2022), no. 6, 2385–2397, DOI 10.5802/aif.3494. MR4500359

[K14] B. Kuckuck, *Subdirect products of groups and the n-$(n+1)$-$(n+2)$ conjecture*, Q. J. Math. **65** (2014), no. 4, 1293–1318, DOI 10.1093/qmath/hau004. MR3285772

[La67] J. P. Labute, *Classification of Demushkin groups*, Canadian J. Math. **19** (1967), 106–132, DOI 10.4153/CJM-1967-007-8. MR210788

[LIMP21] C. Llosa Isenrich, B. Martelli, and P. Py, *Hyperbolic groups containing subgroups of type $\mathcal{F}_3$ not $\mathcal{F}_4$*, J. Differential Geom. **127** (2024), no. 3, 1121–1147, DOI 10.4310/jdg/1721071498. MR4773175

[LIP23] C. Llosa Isenrich, P. Py, *Groups with exotic finiteness properties from complex Morse theory*, Preprint, (2023).

[Lu02] W. Lück, *L^2-invariants: theory and applications to geometry and K-theory*, Ergebnisse der Mathematik und ihrer Grenzgebiete. 3. Folge. A Series of Modern Surveys in Mathematics [Results in Mathematics and Related Areas. 3rd Series. A Series of Modern Surveys in Mathematics], vol. 44, Springer-Verlag, Berlin, 2002, DOI 10.1007/978-3-662-04687-6. MR1926649

[L82] A. Lubotzky, *Combinatorial group theory for pro-p-groups*, J. Pure Appl. Algebra **25** (1982), no. 3, 311–325, DOI 10.1016/0022-4049(82)90086-X. MR666024

[NS2-07] N. Nikolov and D. Segal, *On finitely generated profinite groups. I. Strong completeness and uniform bounds*, Ann. of Math. (2) **165** (2007), no. 1, 171–238, DOI 10.4007/annals.2007.165.171. MR2276769

[NS2-07] N. Nikolov and D. Segal, *On finitely generated profinite groups. II. Products in quasisimple groups*, Ann. of Math. (2) **165** (2007), no. 1, 239–273, DOI 10.4007/annals.2007.165.239. MR2276770

[R88] B. Renz, *Geometrische Invarianten und Endlichkeitseigenschaften von Gruppen*, PhD Thesis, Johann Wolfgang Goethe-Universität Frankfurt am Main, 1988.

[R89] B. Renz, *Geometric invariants and HNN-extensions*, Group theory (Singapore, 1987), de Gruyter, Berlin, 1989, pp. 465–484. MR981863

[RZ10] L. Ribes and P. Zalesskii, *Profinite groups*, 2nd ed., Ergebnisse der Mathematik und ihrer Grenzgebiete. 3. Folge. A Series of Modern Surveys in Mathematics [Results in Mathematics and Related Areas. 3rd Series. A Series of Modern Surveys in Mathematics], vol. 40, Springer-Verlag, Berlin, 2010, DOI 10.1007/978-3-642-01642-4. MR2599132

[R92] V. A. Roman′kov, *Generating elements of groups of automorphisms of free metabelian pro-p-groups* (Russian, with Russian summary), Sibirsk. Mat. Zh. **33** (1992), no. 5,

145–158, 223, DOI 10.1007/BF00970996; English transl., Siberian Math. J. **33** (1992), no. 5, 878–890 (1993). MR1197081

[R93] V. A. Roman′kov, *Infinite generation of groups of automorphisms of free pro-p-groups* (Russian, with English and Russian summaries), Sibirsk. Mat. Zh. **34** (1993), no. 4, 153–159, iv, ix–x, DOI 10.1007/BF00975176; English transl., Siberian Math. J. **34** (1993), no. 4, 727–732. MR1248800

[S71] J.-P. Serre, *Structure de certains pro-p-groupes*, Seminaire Bourbaki 1962/63(252), 357–364 (1971).

[S] J.-P. Serre, *Galois cohomology*, Springer-Verlag, Berlin, 1997. Translated from the French by Patrick Ion and revised by the author, DOI 10.1007/978-3-642-59141-9. MR1466966

[SZ22] I. Snopce and P. Zalesskii, *Right-angled Artin pro-p groups*, Bull. Lond. Math. Soc. **54** (2022), no. 5, 1904–1922. MR4523571

[V23] S. Vidussi, *Higher incoherence of the automorphism groups of free groups*, Preprint, (2023).

[W98] J. S. Wilson, *Profinite groups*, London Mathematical Society Monographs. New Series, vol. 19, The Clarendon Press, Oxford University Press, New York, 1998. MR1691054

[W91] J. S. Wilson, *Finite presentations of pro-p groups and discrete groups*, Invent. Math. **105** (1991), no. 1, 177–183, DOI 10.1007/BF01232262. MR1109625

[Wi19] D. T. Wise, *An invitation to coherent groups*, What's next?—the mathematical legacy of William P. Thurston, Ann. of Math. Stud., vol. 205, Princeton Univ. Press, Princeton, NJ, 2020, pp. 326–414, DOI 10.2307/j.ctvthhdvv.16. MR4205645

[ZZ20] P. Zalesskii and T. Zapata, *Profinite extensions of centralizers and the profinite completion of limit groups*, Rev. Mat. Iberoam. **36** (2020), no. 1, 61–78, DOI 10.4171/rmi/1121. MR4061982

[Za17] M. C. B. Zaremsky, *Separation in the BNSR-invariants of the pure braid groups*, Publ. Mat. **61** (2017), no. 2, 337–362, DOI 10.5565/PUBLMAT6121702. MR3677865

Department of Mathematics, State University of Campinas (UNICAMP), São Paulo, Brasil
Email address: desi@unicamp.br

Department of Mathematics, University of California, Riverside, California 92521
Email address: svidussi@ucr.edu

Contemporary Mathematics
Volume **816**, 2025
https://doi.org/10.1090/conm/816/16351

L^2-type invariants for complex smooth quasi-projective varieties: A survey

Yongqiang Liu

ABSTRACT. Let X be a complex smooth quasi-projective variety with an epimorphism $\nu\colon \pi_1(X) \twoheadrightarrow \mathbb{Z}^n$. We survey recent developments about the asymptotic behaviour of Betti numbers with any field coefficients and the order of the torsion part of singular integral homology of finite abelian covers of X associated to ν, known as the L^2-type invariants. We give relations between L^2-type invariants, Alexander invariants, and cohomology jump loci. When ν is orbifold effective, we give explicit formulas for L^2-invariants at homological degree one in terms of geometric information of X. We also propose several open questions for hyperplane arrangement complement.

1. Introduction

There is a general principle to consider a classical invariant of a finite CW complex and to define its analog for some covering space. This leads to the L^2-type invariants. Atiyah [**Ati76**] introduced the notion of L^2-Betti numbers in the context of a regular covering of a closed Riemannian manifold. After that, there has been vast literature for the L^2-invariant theory; see [**Luc02**]. The L^2-invariant theory has attracted considerable interest and has relations to many other fields, such as operator theory and algebraic K-theory. A particular important result is Lück's approximation theorem [**Luc94**], which states that the L^2-Betti numbers of the universal cover of a finite polyhedron can be found as limits of normalized Betti numbers of finite sheeted normal coverings.

In this paper, we focus on a rather simple case of Lück's approximation problem; see, e.g., Lück's survey papers [**Luc13, Luc16**] on this topic. Let X be a connected finite CW complex with a fixed epimorphism $\nu\colon \pi_1(X) \twoheadrightarrow \mathbb{Z}^n$. Consider a descending chain of subgroups

$$\mathbb{Z}^n = \Gamma_0 \supseteq \Gamma_1 \supseteq \Gamma_2 \supseteq \cdots$$

2020 *Mathematics Subject Classification.* Primary 14F45, 14H30, 32S20; Secondary 32S20.

Key words and phrases. Alexander polynomial, Mahler measure, jump loci, orbifold map, L^2-Betti number, BNS-invariants, hyperplane arrangement.

The author is supported by National Key Research and Development Project SQ2020YFA070080, NSFC grant No. 12001511, the Project of Stable Support for Youth Team in Basic Research Field, CAS (YSBR-001), the project "Analysis and Geometry on Bundles" of Ministry of Science and Technology of the People's Republic of China and Fundamental Research Funds for the Central Universities.

such that Γ_j is a subgroup in $\mathbb{Z}^n$ with finite index and $\bigcap_{j=0}^{\infty} \Gamma_j = \{0\}$. Let X_j denote the finite covering of X associated to the composed epimorphism ν with the quotient map $\mathbb{Z}^n \twoheadrightarrow \mathbb{Z}^n/\Gamma_j$. We study the asymptotic behaviour of the ith Betti number with coefficients in a field coefficient $\mathbb{K}$ and the logarithm of the cardinality of the torsion subgroup of the ith homology group with integral coefficients, i.e., the two sequences

$$\left(\frac{\dim H_i(X_j, \mathbb{K})}{|\mathbb{Z}^n/\Gamma_j|}\right)_{j\geq 0} \quad \text{and} \quad \left(\frac{\log |H_i(X_j, \mathbb{Z})_{\mathrm{tor}}|}{|\mathbb{Z}^n/\Gamma_j|}\right)_{j\geq 0}.$$

Here $|H_i(X_j, \mathbb{Z})_{\mathrm{tor}}|$ denotes the order of the torsion part of $H_i(X_j, \mathbb{Z})$.

One can find the following approximation problem proposed by Lück in [**Luc13**, **Luc16**].

QUESTION 1.1 (Approximation problem).

(i) Does the sequence converge?
(ii) If yes, is the limit independent of the chain?
(iii) If yes, what is the limit?

Luckily for us, Question 1.1 has positive answers in our case as we explain now. For a subgroup $\Gamma \subset \mathbb{Z}^n$ of finite index, we set

$$\langle \Gamma \rangle = \min \left\{ \sum_{j=1}^{n} x_j^2 \mid x = (x_1, \ldots, x_n) \in \Gamma, x \neq 0 \right\}.$$

Let $X^{\nu,\Gamma}$ denote the covering space of X associated to the compositions of ν and the quotient map $\mathbb{Z}^n \to \mathbb{Z}^n/\Gamma$. Consider the following limits:

$$\alpha_i(X^\nu, \mathbb{K}) := \lim_{\langle \Gamma \rangle \to \infty} \frac{\dim H_i(X^{\nu,\Gamma}, \mathbb{K})}{|\mathbb{Z}^n/\Gamma|}$$

for any field coefficients $\mathbb{K}$ and

$$\mathbf{M}_i(X^\nu) := \limsup_{\langle \Gamma \rangle \to \infty} \frac{\log |H_i(X^{\nu,\Gamma}, \mathbb{Z})_{\mathrm{tor}}|}{|\mathbb{Z}^n/\Gamma|}.$$

It is known that the first limit always exists for any field coefficients and does not depend on the choice of chains, see [**LLS11**, Theorem 0.2]. When $\mathrm{char}(\mathbb{K}) = 0$, this follows from the well known Lück's approximation theorem [**Luc94**]. On the other hand, it was proved by Le [**Le14**] that the upper limit $\mathbf{M}_*(X^\nu)$ exists and can be computed by the Mahler measure of the ith integral Alexander polynomial. When $n = 1$, the upper limit in the definition of $\mathbf{M}_*(X^\nu)$ can be replaced by the ordinary limit.

The L^2-type invariants are particular useful for the following Singer–Hopf conjecture. A connected CW complex is said to be aspherical if its universal cover is contractible.

CONJECTURE 1.2 (Singer–Hopf conjecture). *Let X be an aspherical closed manifold with real dimension $2N$. Then*

$$(-1)^N \chi(X) \geq 0.$$

Conjecture 1.2 is still wide open even in the realm of aspherical complex smooth projective varieties. This is particularly surprising as, in the projective case, one can use a large variety of tools coming from algebraic geometry approaches to such conjecture. For the recent developments in this direction, we refer to [**AW23**, **AMW23**, **DL24**, **DL23**, **LMW21**].

Inspired by these developments, we focus on the aforementioned L^2-type invariants in the context of complex smooth quasi-projective varieties. These L^2-type invariants have a deep connection with cohomology jump loci. The cohomology jump loci of a complex smooth quasi-projective variety have been intensively studied in recent years. In particular, the beautiful structure theorem for cohomology jump loci put strong constraints for the homotopy type of complex smooth quasi-projective variety; see Section 3.1 for a brief introduction. We also refer the readers to Budur and Wang's survey paper [**BW17**] for a comprehensive background on this topic.

These results for cohomology jump loci can be used to prove many good properties about the L^2-type invariants. For example, as an application of the structure theorem, we show that the upper limit in the definition of $\mathbf{M}_*(X^\nu)$ can be replaced by the ordinary limit when X is a smooth complex quasi-projective variety (see Corollary 3.2), which is not true for general topological space. This result is new to the author's knowledge.

Since the aforementioned limits for these L^2-type invariants exist, we rephrase Question 1.1(iii) as follows.

QUESTION 1.3. How does one compute these L^2-type invariants in terms of the algebraic geometric information of X when X is a smooth complex quasi-projective variety?

In the case of hyperplane arrangement complement, one can ask if these L^2-type invariants are combinatorially determined. Based on Lück's approximation theorem [**Luc94**], a related question is whether the fundamental group of hyperplane arrangement complement is residually finite?

The purpose of this survey is to provide a gentle introduction to these concepts and results, with an emphasis on applications and open questions. In Section 2, we recall the definition of Alexander type invariants and cohomology jump loci and study their relations with L^2-type invariants. In Section 3, we focus on the case when X is a complex smooth quasi-projective variety. In Section 3.1, we recall the structure theorem about cohomology jump loci and its applications to L^2-type invariants. Section 3.2 is devoted to cohomology jump loci and L^2-type invariants at homological degree one. In Section 4, we focus on hyperplane arrangements. We propose many open questions regarding if these invariants are combinatorially determined. In Section 5, we give proofs for some results claimed in the previous sections.

NOTATIONS.

(1) $\mathbb{K}$ is assumed to be an algebraically closed field in the rest of the paper and char($\mathbb{K}$) denotes the characteristic of the field $\mathbb{K}$.
(2) For any prime number $p \geq 2$, let $\overline{\mathbb{F}}_p$ denote the algebraic closure of the finite field $\mathbb{F}_p$ with p elements.

2. Alexander invariants and cohomology jump loci

Let X be a connected finite CW complex with an epimorphism $\nu\colon \pi_1(X) \twoheadrightarrow \mathbb{Z}^n$. We study L^2-type invariants associated to the pair (X,ν).

2.1. Alexander type invariants. Let X^ν denote the covering space of X associated to $\nu\colon \pi_1(X) \twoheadrightarrow \mathbb{Z}^n$. The group of covering transformations of the covering space X^ν is isomorphic to $\mathbb{Z}^n$ and acts on it. By choosing fixed lifts of the cells of X to X^ν, we obtain a free basis for the cellular chain complex of X^ν as $\mathbb{Z}[\mathbf{t}^{\pm}]$-modules, where $\mathbb{Z}[\mathbf{t}^{\pm}]$ is the multi-variable Laurentiu polynomial ring $\mathbb{Z}[t_1^{\pm},\dots,t_n^{\pm}] = \mathbb{Z}[\mathbb{Z}^n]$. So the cellular chain complex of X^ν, $C_*(X^\nu,\mathbb{Z})$, is a bounded complex of finitely generated free $\mathbb{Z}[\mathbf{t}^{\pm}]$-modules:
(2.1)

$$\cdots \to C_{i+1}(X^\nu,\mathbb{Z}) \overset{\partial_i}{\to} C_i(X^\nu,\mathbb{Z}) \overset{\partial_{i-1}}{\to} C_{i-1}(X^\nu,\mathbb{Z}) \overset{\partial_{i-2}}{\to} \cdots \overset{\partial_0}{\to} C_0(X^\nu,\mathbb{Z}) \to 0.$$

DEFINITION 2.1. The *ith integral Alexander homology* $H_i(X^\nu,\mathbb{Z})$ of the pair (X,ν) is by definition the ith homology of $C_*(X^\nu,\mathbb{Z})$ as complex of $\mathbb{Z}[\mathbf{t}^{\pm}]$-modules. Similarly, for any field $\mathbb{K}$, the *ith Alexander homology with* $\mathbb{K}$*-coefficient* $H_i(X^\nu,\mathbb{K})$ of the pair (X,ν) is by definition the ith homology of $C_*(X^\nu,\mathbb{Z})\otimes_{\mathbb{Z}}\mathbb{K}$ as complex of $\mathbb{K}[\mathbf{t}^{\pm}]$-modules, where $\mathbb{K}[\mathbf{t}^{\pm}]$ denotes the multi-variable Laurent polynomial ring $\mathbb{K}[t_1^{\pm},\dots,t_n^{\pm}]$.

With the above free basis for $C_*(X^\nu,\mathbb{Z})$, ∂_i can be written down as a matrix with entries in $\mathbb{Z}[\mathbf{t}^{\pm}]$. Note that $\mathbb{Z}[\mathbf{t}^{\pm}]$ is a Noetherian UFD. Let $\Delta_i(X^\nu)$ denote the greatest common divisor of all nonzero $(\mathrm{rank}\partial_i \times \mathrm{rank}\partial_i)$-minors of ∂_i. When $\partial_i = 0$, $\Delta_i(X^\nu)=1$ by convention.

DEFINITION 2.2. $\Delta_i(X^\nu)$ is called *the ith integral Alexander polynomial of the pair* (X,ν). Note that $\Delta_i(X^\nu)$ is defined uniquely up to multiplication with a unit in $\mathbb{Z}[\mathbf{t}^{\pm}]$.

REMARK 2.3. The complex $C_*(X^\nu,\mathbb{Z})$ determines $\alpha_*(X^\nu,\mathbb{K})$ and $\mathbf{M}_*(X^\nu)$. In fact, for any $\Gamma\subset\mathbb{Z}^n$ with finite index, consider $\mathbb{Z}[\mathbb{Z}^n/\Gamma]$ as a quotient ring of $\mathbb{Z}[\mathbf{t}^{\pm}]$. Then $H_*(X^{\nu,\Gamma},\mathbb{Z})$ can be computed by the following complex (viewed as a complex of abelian groups):

$$C_*(X^\nu,\mathbb{Z})\otimes_{\mathbb{Z}[\mathbf{t}^{\pm}]}\mathbb{Z}[\mathbb{Z}^n/\Gamma].$$

So it is no surprise that the L^2-type invariants have a deep connection with the Alexander invariants.

Next, we recall the definition of Mahler measure. Let $h\in\mathbb{Z}[\mathbf{t}^{\pm}]$ be a nonzero polynomial. The Mahler measure of h is defined by

$$\mathbf{M}(h) := \int_{(S^1)^n} \log|h(s)|\mathrm{d}s,$$

where $\mathrm{d}s$ indicates the integration with respect to the normalized Haar measure, and $(S^1)^n$ is the multiplicative subgroup of n-dimensional complex space $\mathbb{C}^n$ consisting of all vectors $(s_1,\dots,s_n)$ with $|s_1|=\cdots=|s_n|=1$. When $n=1$ Mahler measure has the following equivalent definition.

DEFINITION 2.4 ([**Mah60, Mah62**]). For a nonzero polynomial $h(t)\in\mathbb{Z}[\mathbf{t}^{\pm}]$ with $n=1$, assume that $h(t)=a_dt^d+a_{d-1}t^{d-1}+\cdots+a_0 = a_d\cdot\prod_{j=1}^d(t-\alpha_j)$ with $a_d\neq 0$, $a_0\neq 0$ and $\alpha_j\in\mathbb{C}$ for all $1\le j\le d$.

We define the *Mahler measure* of h to be

$$\mathbf{M}(h) = \log(|a_d| \cdot \prod_{j=1}^{d} \max\{1, |\alpha_j|\}).$$

Recall that an element $h \in \mathbb{Z}[\mathbf{t}^{\pm}]$ is called a generalized cyclotomic polynomial [**Sch15**, p. 47] if it is of the form $h(t_1, \dots, t_n) = \Phi(\prod_{i=1}^{n} t_i^{a_i})$ (up to units in $\mathbb{Z}[\mathbf{t}^{\pm}]$), where $0 \neq (a_1, \cdots, a_n) \in \mathbb{Z}^n$ and Φ is a cyclotomic polynomial in a single variable. The following theorem due to Schmidt characterizes when the Mahler measure could be zero.

THEOREM 2.5 ([**Sch95**, Theorem 19.5]). *For $h \in \mathbb{Z}[\mathbf{t}^{\pm}]$, $\mathbf{M}(h) = 0$ if and only if h is a product of generalized cyclotomic polynomials up to units.*

Based on the work of Schmidt [**Sch95**, Theorem 21.1], Le showed that [**Le14**, Theorem 5] $\mathbf{M}_i(X^\nu)$ can be computed by the Mahler measure of $\Delta_i(X^\nu)$. When $n = 1$, the following result is also proved by Raimbault [**Rai12**, Theorem 0.2] independently.

THEOREM 2.6 ([**Le14**, Theorem 5]). *Let X be a connected finite CW complex with an epimorphism $\nu\colon \pi_1(X) \twoheadrightarrow \mathbb{Z}^n$. Then for any i we have*

$$\mathbf{M}_i(X^\nu) = \mathbf{M}(\Delta_i(X^\nu)).$$

If $n = 1$ then lim sup *in the definition of $\mathbf{M}_i(X^\nu)$ can be replaced by the ordinary* lim.

Inspired by Theorem 2.5, we prove the following proposition.

PROPOSITION 2.7. *Let X be a connected finite CW complex with an epimorphism $\nu\colon \pi_1(X) \twoheadrightarrow \mathbb{Z}^n$. If $\Delta_i(X^\nu)$ is a product of some generalized cyclotomic polynomials with a nonzero integer c_i, then* lim sup *in the definition of $\mathbf{M}_i(X^\nu)$ can be replaced by the ordinary* lim *and $\mathbf{M}_i(X^\nu) = \log|c_i|$.*

The proof of this proposition is given in Section 5.

2.2. Cohomology jump loci. We first recall the definition of cohomology jump loci. For any algebraically closed field $\mathbb{K}$, the group of $\mathbb{K}$-valued characters, $\mathrm{Hom}(\pi_1(X), \mathbb{K}^*)$, is a commutative affine algebraic group. Each character ρ in $\mathrm{Hom}(\pi_1(X), \mathbb{K}^*)$ defines a rank one local system on X, denoted by L_ρ. Since $\mathbb{K}^*$ is abelian, $\mathrm{Hom}(\pi_1(X), \mathbb{K}^*)$ in fact only depends on $H_1(X, \mathbb{Z})$, the abelianization of $\pi_1(X)$, and $\mathrm{Hom}(\pi_1(X), \mathbb{K}^*) = \mathrm{Hom}(H_1(X, \mathbb{Z}), \mathbb{K}^*)$.

DEFINITION 2.8. The cohomology jump loci of X are defined as

$$\mathcal{V}^i_k(X, \mathbb{K}) := \{\rho \in \mathrm{Hom}(\pi_1(X), \mathbb{K}^*) \mid \dim_{\mathbb{K}} H^i(X, L_\rho) \geq k\}.$$

Cohomology jump loci are closed subvarieties of $\mathrm{Hom}(\pi_1(X), \mathbb{K}^*)$ and homotopy invariants of X. In cohomological degree one, $\mathcal{V}^1_k(X, \mathbb{K})$ depends only on $\pi_1(X)$ (e.g., see [**Suc11**, Section 2.2]).

The map $\nu\colon \pi_1(X) \twoheadrightarrow \mathbb{Z}^n$ induces an embedding $\nu^*\colon (\mathbb{K}^*)^n \hookrightarrow \mathrm{Hom}(\pi_1(X), \mathbb{K}^*)$. For a tuple $\lambda = (\lambda_1, \dots, \lambda_n) \in (\mathbb{K}^*)^n$, let $\nu^{-1}L_\lambda$ denote the corresponding rank one $\mathbb{K}$-local system on X whose monodromy representation factors through ν.

PROPOSITION 2.9. *With the notations and assumptions as above, for any* $i \geq 0$ *and* $\lambda \in (\mathbb{K}^*)^n$ *being general we have*

$$\alpha_i(X^\nu, \mathbb{K}) = \dim H_i(X, \nu^{-1}L_\lambda) = \mathrm{rank} H_i(X^\nu, \mathbb{K}).$$

Here $\mathrm{rank} H_i(X^\nu, \mathbb{K})$ *is the dimension of the vector space* $H_i(X^\nu, \mathbb{K}) \otimes \mathcal{F}$ *over the fractional field* $\mathcal{F}$ *of* $\mathbb{K}[\mathbf{t}^{\pm}]$. *In particular,* $\alpha_i(X^\nu, \mathbb{K})$ *is always an integer. Moreover, if* $\Delta_i(X^\nu)$ *is not a unit in* $\mathbb{C}[\mathbf{t}^{\pm}]$, *every irreducible component of the zero locus of* $\Delta_i(X^\nu)$ *in* $(\mathbb{C}^*)^n$ *is a irreducible component of* $\mathcal{V}^i_k(X, \mathbb{C}) \cap (\mathbb{C}^*)^n$ *(via* ν^**) with codimension one for* $k = \mathrm{rank} H_i(X^\nu, \mathbb{C}) + 1$.

This proposition is not new and the claims are either already proved in [**LL23**, Propsition 2.2] or folklore. We give a brief proof in Section 5 due to the lack of precise references. For more relations between cohomology jump loci and Alexander type invariants, we refer the readers to Suciu's paper [**Suc11**].

We also list some folklore properties of L^2-type invariants, which would be useful later.

PROPOSITION 2.10. *Let* X *be a connected finite CW complex with a fixed epimorphism* $\nu\colon \pi_1(X) \twoheadrightarrow \mathbb{Z}^n$. *Then we have the following properties.*

(i) $\alpha_*(X^\nu, \mathbb{K})$ *and* $\mathbf{M}_*(X^\nu)$ *are homotopy invariant of the pair* (X, ν).

(ii) *Euler number: For any field* $\mathbb{K}$,

$$\sum_{i \geq 0} (-1)^i \alpha_i(X^\nu, \mathbb{K}) = \chi(X).$$

(iii) *Poincare duality: If* X *is an oriented compact manifold with* $\dim_{\mathbb{R}} X = N$, *then*

$$\alpha_i(X^\nu, \mathbb{K}) = \alpha_{N-i}(X^\nu, \mathbb{K}) \text{ and } \mathbf{M}_i(X^\nu) = \mathbf{M}_{N-i-1}(X^\nu).$$

(iv) *Zeroth* L^2*-type invariants:* $\alpha_0(X^\nu, \mathbb{K}) = 0$ *and* $\mathbf{M}_0(X^\nu) = 0$.

3. Complex smooth quasi-projective variety

In this section, we always assume that X is a complex smooth quasi-projective variety.

3.1. Structure theorem. The following structure theorem for $\mathcal{V}^i_k(X, \mathbb{C})$ put strong constraints for the homotopy type of complex smooth quasi-projective variety. Many people contributed and we name a few here: Green and Lazarsfeld [**GL91**], Simpson [**Sim93A**], Arapura [**Ara97**], Dimca, Papadima, and Suciu [**DPS09**], Artal Bartolo, Cogolludo-Agustín, and Matei [**ACM13**], Dimca and Papadima [**DP14**], Schnell [**Sch15**], etc. It is finalized by Budur and Wang in [**BW15, BW20**].

THEOREM 3.1 ([**BW15, BW20**]). *Let* X *be a complex smooth quasi-projective variety. Then for any* i *and* k, $\mathcal{V}^i_k(X, \mathbb{C})$ *is a finite union of torsion translated sub-tori of* $\mathrm{Hom}(\pi_1(X), \mathbb{C}^*)$. *Moreover, for every positive dimensional irreducible component* V *of* $\mathcal{V}^i_k(X, \mathbb{C})$, *there exists a torsion point* $\rho \in \mathrm{Hom}(\pi_1(X), \mathbb{C}^*)$ *and an algebraic map* $f\colon X \to G$ *with* G *being a complex semi-abelian variety such that*

$$V = \rho \cdot \mathrm{im}\{f^*\colon \mathrm{Hom}(\pi_1(G), \mathbb{C}^*) \hookrightarrow \mathrm{Hom}(\pi_1(X), \mathbb{C}^*)\}.$$

If X *is a smooth projective variety,* G *has to be an abelian variety, hence every irreducible component of* $\mathcal{V}^i_k(X, \mathbb{C})$ *is even dimensional.*

As a direct application of the above structure theorem, Theorem 2.6, Proposition 2.7 and Proposition 2.9, we have the following results.

COROLLARY 3.2. *Let X be a complex smooth quasi-projective variety with an epimorphism $\nu\colon \pi_1(X) \to \mathbb{Z}^n$. Then for any i, $\Delta_i(X^\nu)$ is product of some generalized cyclotomic polynomials and a nonzero integer c_i. Then* lim sup *in the definition of $\mathbf{M}_i(X^\nu)$ can be replaced by the ordinary* lim *and we have*

$$\mathbf{M}_i(X^\nu) = \log |c_i|.$$

Let X be a complex smooth projective variety. For any subgroup $\Gamma \subset \mathbb{Z}^n$ with finite index, $X^{\nu,\Gamma}$ is also a complex smooth projective variety. In particular, $H^i(X^{\nu,\Gamma}, \mathbb{Q})$ has pure Hodge structures. Instead of studying the asymptotic behavior of Betti number of $H^i(X^{\nu,\Gamma}, \mathbb{Q})$, one can also study the asymptotic behavior of the Hodge numbers of $H^i(X^{\nu,\Gamma}, \mathbb{Q})$. Let $h^{p,q}(X^{\nu,\Gamma})$ denote the dimension of $H^q(X^{\nu,\Gamma}, \Omega^p_{X^{\nu,\Gamma}})$.

PROPOSITION 3.3. *With the above notations and assumptions, for any (p,q) we have*

$$\lim_{\langle\Gamma\rangle\to\infty} \frac{h^{p,q}(X^{\nu,\Gamma})}{|\mathbb{Z}^n/\Gamma|} = \dim H^q(X, \Omega^p_X \otimes \mathcal{L}'),$$

where $\mathcal{L}' = L' \otimes \mathcal{O}_X$ and L' is a generic unitary rank one local system in $\mathrm{im}\nu^*$. *Here ν^* is the embedding $(\mathbb{C}^*)^n \hookrightarrow \mathrm{Hom}(\pi_1(X), \mathbb{C}^*)$ induced by ν.*

The proof of this proposition is given in Section 5. When ν is the natural map to $H_1(X, \mathbb{Z})$ modulo torsions and the Albanese map of X is semi-small, Di Cerbo and Lombardi [**DL24**] give an effective estimate for the asymptotic behaviour of the Hodge numbers and showed that

$$\lim_{\langle\Gamma\rangle\to\infty} \frac{h^{p,q}(X^{\nu,\Gamma})}{|\mathbb{Z}^n/\Gamma|} = \begin{cases} (-1)^q \chi(X, \Omega^p_X), & \text{if } p+q = \dim_{\mathbb{C}} X, \\ 0, & \text{else.} \end{cases}$$

Using the generic vanishing results proved by Popa and Schnell [**PS13**], Di Cerbo and Lombardi's results are compatible with Proposition 3.3.

One may ask about Proposition 3.3 for smooth quasi-projective varieties. In this case, $H^i(X^{\nu,\Gamma}, \mathbb{Q})$ has mixed Hodge structures.

QUESTION 3.4. Doe the limit for the Hodge numbers of $H^i(X^{\nu,\Gamma}, \mathbb{Q})$ exist?

Budur proved [**Bud09**, Theorems 1.8, 1.9] the polynomial periodicity for certain Hodge numbers of congruence covers of complex smooth quasi-projective variety based on a structure theorem for line bundles [**Bud09**, Theorems 1.3, 1.4]. His work shed light on the above question at least for the congruence covers. For more references about polynomial periodicity of congruence covers, we refer the readers to Suciu's survey paper [**Suc01**, Section 5].

3.2. Degree one case.

3.2.1. *Cohomology jump loci.* If we focus on cohomological degree one jump loci, a celebrity result due to Beauville [**Bea92**], Arapura [**Ara97**], and Artal Bartolo, Cogolludo-Agustín, and Matei [**ACM13**] puts even stronger constraints on its fundamental group. To explain their results, we introduce the notion of orbifold maps.

DEFINITION 3.5. Let X be a smooth complex quasi-projective variety. An algebraic map $f\colon X \to C_{g,r}$ is called an orbifold map, if f is surjective, has connected generic fiber and $C_{g,r}$ is a smooth algebraic curve of genus g with r points removed. We always assume that $C_{g,r} \neq \mathbb{CP}^1, \mathbb{C}$. There exists a maximal Zariski open subset $U \subset C_{g,r}$ such that f is a fibration over U. Say $B = C_{g,r} - U$ (could be empty) has s points, denoted by $\{q_1, \ldots, q_s\}$. We assign the multiplicity μ_j of the fiber $f^*(q_j)$ (the gcd of the coefficients of the divisor f^*q_j) to the point q_j. Such orbifold map f is called of type $(g, r, \boldsymbol{\mu})$, where $\boldsymbol{\mu} = (\mu_1, \ldots, \mu_s)$. When $B = \emptyset$, $\prod_{j=1}^{s} \mu_j = 1$ by convention.

When $C_{g,r}$ is clear in the context, we simply write C. Our definition of orbifold maps is a bit of different from the classical one; see, e.g., [**DS14**, Section 4.1]. The classical one will be called the hyperbolic orbifold map, where we adopt the name from [**Del08**].

DEFINITION 3.6. An orbifold map $f\colon X \to C$ of type $(g, r, \boldsymbol{\mu})$ is called hyperbolic if

$$2g + r - 2 + \sum_{j=1}^{s}(1 - \frac{1}{\mu_j}) > 0.$$

This definition is used to rule out the case where C is either $\mathbb{C}^*$ or an elliptic curve and f has no multiple fibers. The orbifold group $\pi_1^{\mathrm{orb}}(C_{g,r}, \boldsymbol{\mu})$ associated to these data is defined as

$$\pi_1^{\mathrm{orb}}(C_{g,r}, \boldsymbol{\mu}) := \pi_1(C_{g,r} \backslash \{q_1, \ldots, q_s\}) / \langle \gamma_j^{\mu_j} = 1 \text{ for all } 1 \leq j \leq s \rangle,$$

where γ_j is a meridian of q_j. An orbifold map $f\colon X \to C$ of type $(g, r, \boldsymbol{\mu})$ induces a surjective map to the orbifold group (see [**ACM10**, Proposition 1.4])

$$f_*\colon \pi_1(X) \twoheadrightarrow \pi_1^{\mathrm{orb}}(C_{g,r}, \boldsymbol{\mu}).$$

Hence it induces an embedding for any $k \geq 1$

$$\mathcal{V}_k^1(\pi_1^{orb}(C_{g,r}, \boldsymbol{\mu}), \mathbb{K}) \to \mathcal{V}_k^1(X, \mathbb{K}).$$

The following theorem shows that every positive-dimensional component of $\mathcal{V}_1^1(X, \mathbb{C})$ is characterized by the orbifold map in this way. The idea first appeared in Beauville's work [**Bea92**] for the projective case and was extended to the quasi-projective case by Arapura [**Ara97**] and Artal Bartolo, Cogolludo-Agustín, and Matei [**ACM13**]. Further properties were found by Dimca [**Dim07, Dim09**], Dimca, Papadima, and Suciu [**DPS09**], etc.

THEOREM 3.7 ([**Ara97, ACM13, Dim07**]). *Let X be a complex smooth quasi-projective variety.*

(a) *We have*

$$\mathcal{V}_1^1(X, \mathbb{C}) = \bigcup_f f^* \mathcal{V}_1^1(\pi_1^{\mathrm{orb}}(C_{g,r}, \boldsymbol{\mu})), \mathbb{C}) \cup Z,$$

where Z is a finite set of torsion points and the union runs over all hyperbolic orbifold maps f.

(b) *Fix an orbifold map $f\colon X \to C$ of type $(g, r, \boldsymbol{\mu})$. Then, for any*

$$\rho \in \mathrm{Hom}(\pi_1^{\mathrm{orb}}(C_{g,r}, \boldsymbol{\mu}), \mathbb{C}^*)$$

we have

$$\dim H^1(X, f^*L_\rho) \geq \dim H^1(\pi_1^{\text{orb}}(C_{g,r}, \boldsymbol{\mu}), L_\rho)$$

and equality holds with finitely many (torsion) exceptions.

REMARK 3.8. This theorem indeed gives a complete description of all positive-dimensional components contained in $\mathcal{V}_k^1(X, \mathbb{C})$ for all $k \geq 1$.

REMARK 3.9. In Theorem 3.7(a), one can indeed take the union with a further restriction that if $g = 1$ and $r = 0$, $\dfrac{\prod_{j=1}^s \mu_j}{\operatorname{lcm}(\mu_1, \cdots, \mu_s)} > 1$, since otherwise $\mathcal{V}_1^1(\pi_1^{\text{orb}}(C_{1,0}, \boldsymbol{\mu})), \mathbb{C}) = \{1\}$ (see, e.g., [**Suc21**, Section 10]).

When $\operatorname{char}(\mathbb{K}) = p > 0$, not much is known for $\mathcal{V}_k^i(X, \mathbb{K})$. So far, all the known results concern cohomological degree one jump loci. Delzant [**Del08**] enabled the generalization of Theorem 3.7(a) to any algebraically closed $\mathbb{K}$-coefficient case when X is a complex smooth projective variety.

THEOREM 3.10 ([**Del08**]). *Let X be a complex smooth projective variety and $\mathbb{K}$ be an algebraically closed field. Then we have*

$$\mathcal{V}_1^1(X, \mathbb{K}) = \bigcup_f f^*\mathcal{V}_1^1(\pi_1^{\text{orb}}(\Sigma_{g,r}, \boldsymbol{\mu})), \mathbb{K}) \cup Z',$$

where Z' is a finite set of torsion points (depending on $\mathbb{K}$) and the union runs over all hyperbolic orbifold maps f. In particular, the structure theorem holds for $\mathcal{V}_1^1(X, \mathbb{K})$.

The following theorem due to Li and the author generalizes Theorem 3.7(b) to any algebraically closed field coefficients.

THEOREM 3.11 ([**LL24**, Theorem 1.6]). *Let X be a complex smooth quasi-projective variety and $\mathbb{K}$ be an algebraically closed field. Given an orbifold map $f\colon X \to C$ of type $(g, r, \boldsymbol{\mu})$ and any $\rho \in \operatorname{Hom}(\pi_1^{\text{orb}}(C_{g,r}, \boldsymbol{\mu}), \mathbb{K}^*)$, we have*

$$\dim H^1(X, f^*L_\rho) \geq \dim H^1(\pi_1^{\text{orb}}(C_{g,r}, \boldsymbol{\mu}), L_\rho)$$

and equality holds with finitely many (torsion) exceptions.

REMARK 3.12. As in Remark 3.8, Theorem 3.10 and Theorem 3.11 together give a complete description for all positive-dimensional components contained in $\mathcal{V}_k^1(X, \mathbb{K})$ for all k when X is smooth projective. Hence the structure theorem holds for $\mathcal{V}_k^1(X, \mathbb{K})$ for all $k \geq 1$.

3.2.2. *BNS-invariants.* One may wonder about Theorem 3.10 for smooth quasi-projective varieties. Following Delzant's approach in [**Del08, Del10**], one needs to study the BNS-invariants of X. Let us recall its definition.

Pick a finite generating set of $\pi_1(X)$ and let $\Omega(X)$ be the corresponding Cayley graph of $\pi_1(X)$. Let $\mathrm{S}(X)$ denote the unit sphere in the real vector space $\operatorname{Hom}(\pi_1(X); \mathbb{R})$. For any character $\chi \in \mathrm{S}(X)$, let $\Omega_\chi(X)$ be the full subgraph on vertex set $\{g \in \pi_1(X) \mid \chi(g) \geq 0\}$. Bieri, Neumann, and Strebel defined [**BNS87**] a subset $\Sigma^1(X)$ of $H^1(X, \mathbb{R}) \setminus \{0\}$, nowadays called the BNS-invariant of X as follows; see, e.g., [**PS10**, Definition 7.1].

DEFINITION 3.13. The set $\Sigma^1(X)$ consists of those $\chi \in \mathrm{S}(X)$ for which the graph $\Omega_\chi(X)$ is connected.

Papadima and Suciu studied the relations between $\Sigma^1(X)$ and the orbifold maps in [**PS10**, Section 16], and the results were recently refined by Suciu via the tropical method [**Suc21**].

THEOREM 3.14. *Let X be a complex smooth quasi-projective variety. Then*

$$\Sigma^1(X) \subseteq \mathrm{S}\Big(\bigcup_f \mathrm{im}(f^*\colon H^1(C,\mathbb{R}) \to H^1(X,\mathbb{R}))\Big)^c,$$

where the finite union runs over all hyperbolic orbifold maps f. Here $\mathrm{S}()^c$ means taking the complement of the projection of the set to the character sphere.

In fact, a result of Delzant [**Del10**] shows that the above inclusion holds as an equality when X is a smooth projective variety (or even a compact Kähler manifold). Based on a deep theorem due to Simpson [**Sim93B**], Delzant gave a complete description of BNS-invariants of complex compact Kähler manifold [**Del10**], which are characterized by the hyperbolic orbifold maps. Using this description of BNS-invariants by orbifold maps, Delzant managed to prove Theorem 3.10 in [**Del08**].

QUESTION 3.15. Inspired by these works, one may ask the following questions:

(i) Does [**Sim93B**, Theorem 1] hold for complex smooth quasi-projective varieties?
(ii) Does the inclusion in Theorem 3.14 hold as an equality?
(iii) Does Theorem 3.10 hold for complex smooth quasi-projective varieties?

Ideally a positive answer to Question 3.15(i) (resp., (ii)) would imply a positive answer to Question 3.15(ii) (resp., (iii)). Question 3.15(ii) was first raised by Papadima and Suciu in [**PS10**] and refined by Suciu in [**Suc21**, Question 13.7] for hyperplane arrangement complements.

3.2.3. *L^2-type invariants.* To compute the L^2-type invariants of the pair (X,ν) at homological degree one, we need to connect ν with the orbifold maps.

DEFINITION 3.16. Let X be a smooth complex quasi-projective variety with an epimorphism $\nu\colon \pi_1(X) \twoheadrightarrow \mathbb{Z}^n$. We say that ν is orbifold effective if there is an orbifold map $f\colon X \to C$ of type $(g,r,\boldsymbol{\mu})$ such that ν factors through f_* as follows:

$$\begin{array}{ccc} \pi_1(X) & \xrightarrow{\quad\nu\quad} & \mathbb{Z}^n . \\ & {\scriptstyle f_*}\searrow \qquad \nearrow & \\ & \pi_1^{\mathrm{orb}}(\Sigma_{g,r},\boldsymbol{\mu}) & \end{array}$$

We say that ν is orbifold effective by f and call ν being of type $(g,r,\boldsymbol{\mu})$.

REMARK 3.17. As shown in [**LL24**, Remark 4.3], if ν is orbifold effective by two different orbifold maps f and g, then f and g must have the same type $(g,r,\boldsymbol{\mu})$. So it is well defined to call ν being of type $(g,r,\boldsymbol{\mu})$.

The following theorem by Li and the author gave concrete formulas for the L^2-type invariants of X at homological degree one when ν is orbifold effective.

THEOREM 3.18 ([**LL24**, Theorem 1.10], [**LL23**, Theorem 1.3]). *Let X be a complex smooth quasi-projective variety with an epimorphism $\nu\colon \pi_1(X) \twoheadrightarrow \mathbb{Z}^n$. Suppose that ν is orbifold effective of type $(g,r,\boldsymbol{\mu})$. Let $\mathbb{K}$ be a field with* $\mathrm{char}(\mathbb{K}) = p \geq 0$. *Then*

$$\alpha_1(X^\nu,\mathbb{K}) = 2g + r - 2 + \#\{j \mid p \textit{ divides } \mu_j\}$$

and

$$\mathbf{M}_1(X^\nu) = \sum_{j=1}^{s} \log \mu_j.$$

On the other hand, assume that either X is projective or $H^1(X, \mathbb{Q})$ has pure Hodge structure of type $(1,1)$. Then $\alpha_1(X^\nu, \mathbb{K}) > 0$ or $\mathbf{M}_1(X^\nu) > 0$ happens only if ν is orbifold effective.

REMARK 3.19. A positive answer to Question 3.15(iii) would enable us to drop the assumption that either X is projective or $H^1(X, \mathbb{Q})$ has pure Hodge structure of type $(1,1)$ in the last claim of the above theorem.

REMARK 3.20. Due to Proposition 2.9 and Corollary 3.2, the above theorem also implies the corresponding formulas for the rank of the Alexander homology and the leading coefficient of the integral Alexander polynomial at homological degree one.

The above theorem is just about L^2-type invariants at homological degree one. How about higher homological degrees? Following the structure theorem, we propose the following higher homological degree analogs. Let V be a positive-dimensional irreducible component in $\mathcal{V}^i(X, \mathbb{C})$. According to the structure theorem there exists an algebraic map $f\colon X \to G$ with G being a complex semi-abelian variety such that V is contained in a translate of

$$\operatorname{im}\{f^*\colon \operatorname{Hom}(\pi_1(G), \mathbb{C}^*) \hookrightarrow \operatorname{Hom}(\pi_1(X), \mathbb{C}^*)\}.$$

QUESTION 3.21. Note that $\pi_1(G)$ is a finitely generated free abelian group. Consider the epimorphism $\nu = f_*\colon \pi_1(X) \to \pi_1(G)$ induced by f. Can one compute $\alpha_i(X^\nu, \mathbb{K})$ and $\mathbf{M}_i(X^\nu)$ by the geometric information of f? One may need to take the Stein factorization of f.

Using Theorem 3.18 and Proposition 2.10, we give a complete description of the L^2-type invariants when X is a complex smooth projective surface.

PROPOSITION 3.22. *Let X be a complex smooth projective surface with an epimorphism $\nu\colon \pi_1(X) \twoheadrightarrow \mathbb{Z}^n$. Let $\mathbb{K}$ be a field with $\operatorname{char}(\mathbb{K}) = p \geq 0$. If ν is not orbifold effective, then*

$$\alpha_i(X^\nu, \mathbb{K}) = \begin{cases} \chi(X), & \text{if } i = 2, \\ 0, & \text{else} \end{cases}$$

and

$$\mathbf{M}_i(X^\nu) = 0 \text{ for all } i.$$

On other hand, if ν is orbifold effective, say, of type $(g, r, \boldsymbol{\mu})$, then

$$\alpha_i(X^\nu, \mathbb{K}) = \begin{cases} 2g + r - 2 + \#\{j \mid p \text{ divides } \mu_j\}, & \text{if } i = 1, 3, \\ \chi(X) + 4g + 2r - 4 + 2\#\{j \mid p \text{ divides } \mu_j\}, & \text{if } i = 2, \\ 0 & \text{else} \end{cases}$$

and

$$\mathbf{M}_i(X^\nu) = \begin{cases} \sum_{j=1}^{s} \log \mu_j, & \text{if } i = 1, 2, \\ 0, & \text{else.} \end{cases}$$

4. Hyperplane arrangement

In this section, we always assume that X is the complement of a hyperplane arrangement $\mathcal{A}$ in $\mathbb{CP}^N$ for some N and we focus on a fixed epimorphism $\nu\colon \pi_1(X) \to \mathbb{Z}$ for $n = 1$.

For any field $\mathbb{K}$, we denote $\nu_\mathbb{Z}$ and $\nu_\mathbb{K}$ the corresponding element in $H^1(X, \mathbb{Z})$ and $H^1(X, \mathbb{K})$, respectively. Note that $\nu_\mathbb{Z} \cup \nu_\mathbb{Z} = 0$, hence $\nu_\mathbb{K} \cup \nu_\mathbb{K} = 0$. Then we get two chain complexes by cup product

$$(H^*(X,\mathbb{Z}), \cdot\nu_\mathbb{Z})\colon \qquad H^0(X,\mathbb{Z}) \xrightarrow{\nu_\mathbb{Z}} H^1(X,\mathbb{Z}) \xrightarrow{\nu_\mathbb{Z}} H^2(X,\mathbb{Z}) \longrightarrow \cdots$$

$$(H^*(X,\mathbb{K}), \cdot\nu_\mathbb{K})\colon \qquad H^0(X,\mathbb{K}) \xrightarrow{\nu_\mathbb{K}} H^1(X,\mathbb{K}) \xrightarrow{\nu_\mathbb{K}} H^2(X,\mathbb{K}) \longrightarrow \cdots,$$

which are called *the Aomoto complex* of $H^*(X, \mathbb{Z})$ and $H^*(X, \mathbb{K})$ associated to $\nu_\mathbb{Z}$ and $\nu_\mathbb{K}$, respectively.

DEFINITION 4.1. With the above notations, we define *the ith Aomoto Betti number with $\mathbb{K}$-coefficients* as

$$\beta_i(X, \nu_\mathbb{K}) := \dim_\mathbb{K} H^i(H^*(X,\mathbb{K}), \cdot\nu_\mathbb{K}).$$

and *the order of the ith Aomoto torsion* as

$$\tau_i(X, \nu_\mathbb{Z}) := |H^{i+1}(H^*(X,\mathbb{Z}), \cdot\nu_\mathbb{Z})_{\text{tor}}|.$$

Here the shift by 1 is due to the universal coefficient theorem.

As shown by Orlik and Solomon [**OS80**], the cohomology ring $H^*(X, \mathbb{Z})$ are determined by the combinatorics data of the hyperplane arrangement. Then $\beta_*(X, \nu_\mathbb{K})$ and $\tau_*(X, \nu_\mathbb{Z})$ are combinatorially determined, once ν, considered as an element in $H^1(X, \mathbb{Z})$, is fixed. Papadima and Suciu discovered the following results in [**PS10**].

THEOREM 4.2. *With the above notations and assumptions, for any field $\mathbb{K}$ we have*

$$\alpha_i(X^\nu, \mathbb{K}) \leq \beta_i(X, \nu_\mathbb{K}).$$

Moreover, when $\mathbb{K} = \mathbb{C}$, the above inequality holds as a equality.

Papadima and Suciu [**PS10**] indeed proved the above inequality for any finite CW complex. When $\mathbb{K} = \mathbb{C}$, the equality was first proved (using different methods) by Cohen and Suciu [**CS99**], Libgober [**Lib02**], and Libgober and Yuzvinsky [**LY00**] for $i = 1$ and was generalized to the higher degrees by Cohen and Orlik [**CO00**, Theorem 3.7].

On the other hand, the author, jointly with Li, gave the following combinatorial upper bound for $\mathbf{M}_*(X^\nu)$.

PROPOSITION 4.3 ([**LL24**, Proposition 1.12]). *Let X be the complement of a hyperplane arrangement. For any epimorphism $\nu\colon \pi_1(X) \to \mathbb{Z}$ and any $i \geq 0$, we have*

$$\exp(\mathbf{M}_i(X^\nu)) \mid \tau_i(X, \nu_\mathbb{Z}).$$

In particular, if ν is orbifold effective of type $(g, r, \boldsymbol{\mu})$, then

$$\prod_{j=1}^{s} \mu_j \mid \tau_1(X, \nu_\mathbb{Z}). \tag{4.1}$$

In fact, due to Hodge theory, if ν is orbifold effective of type $(g, r, \boldsymbol{\mu})$ for the hyperplane complement X, then g has to be 0. The proof of this proposition is based on Papadima and Suciu's work [**PS10**, Theorem 12.6] and two facts about hyperplane arrangement complement:

- the hyperplane arrangement complement X is homotopy equivalent to a minimal CW complex; see [**DP03, Ran02**].
- $\alpha_i(X^\nu, \mathbb{C}) = \beta_i(X, \nu_\mathbb{C})$ for any i as in Theorem 4.2.

There are examples of hyperplane arrangements where $\alpha_i(X, \mathbb{K}) \neq \beta_i(X, \mathbb{K})$ and $\prod_{j=1}^{s} \mu_j \neq \tau_1(X, \nu_\mathbb{Z})$, see [**LL24**, Example 5.3] and the next example.

EXAMPLE 4.4. Let $\mathcal{B}$ be the deleted B_3-arrangement in $\mathbb{CP}^2$ with defining equations
$$zxy(x-y)(x-z)(y-z)(x-y-z)(x-y+z).$$
Order the hyperplanes as the factors of the defining polynomial. Let X be the complement of the arrangement. It was first discovered by Suciu [**Suc02**] that $\mathcal{V}_1^1(X, \mathbb{C})$ has a translated component
$$V := \rho \otimes \{t, t^{-1}, 1, t^{-1}, t, t^2, t^{-2} \mid t \in \mathbb{C}^*\}$$
with $\rho = (1, -1, -1, -1, 1, 1, 1) \in (\mathbb{C}^*)^7$; see, e.g., [**Dim17**, Example 6.16]. Here we take $\{z = 0\}$ as the hyperplane at infinity. This component is induced by the orbifold map $f\colon X \to \mathbb{C}^*$ as following
$$f([x, y, z]) = \frac{x(y-z)(x-y-z)^2}{y(x-z)(x-y+z)^2}$$
and f is of type $(0, 2, 2)$. Consider $\nu = (1, -1, 0, -1, 1, 2, -2) \in H^1(X, \mathbb{Z})$ induced by f. Then
$$\alpha_1(X^\nu, \mathbb{K}) = \begin{cases} 1, & \text{if } \operatorname{char}(\mathbb{K}) = 2, \\ 0, & \text{if } \operatorname{char}(\mathbb{K}) \neq 2, \end{cases} \quad \text{and } \mathbf{M}_1(X^\nu) = \log 2.$$

On the other hand, the Aomoto complex for $\nu_\mathbb{Z}$ and $\nu_\mathbb{K}$ can be computed by the formula given in [**Dim17**, p.119]. A direct computation shows that
$$\tau_1(X, \nu_\mathbb{Z}) = 4 \text{ and } \beta_1(X, \nu_\mathbb{K}) = \begin{cases} 1, & \text{if } \operatorname{char}(\mathbb{K}) = 2, \\ 0, & \text{if } \operatorname{char}(\mathbb{K}) \neq 2. \end{cases}$$

It is a long-standing open question for hyperplane arrangement if $\mathcal{V}_1^1(X, \mathbb{C})$ is combinatorially determined; see, e.g., [**Suc14**, Problem 3.15]. It is known that the positive-dimensional components of $\mathcal{V}_1^1(X, \mathbb{C})$ passing through the origin are combinatorially determined (see, e.g., [**Dim17**, Theorem 6.1]) and they can be described by the multinet structure, for which we refer, in particular, to the work of Falk and Yuzvinsky [**FY07**]. Since $\mathcal{V}_1^1(X, \mathbb{C})$ only depends on the the fundamental group of X, using the Lefschetz hyperplane section theorem, from now on we assume that X is the complement of a line arrangement in $\mathbb{CP}^2$. Let us now recall the definition of multinet.

DEFINITION 4.5. Let $\mathcal{A}$ be an arrangement in $\mathbb{CP}^2$. A multinet on $\mathcal{A}$ is a partition of $\mathcal{A}$ into $k \geq 3$ subsets $\mathcal{A}_1, \cdots, \mathcal{A}_k$, together with an assignment of multiplicities, $m\colon \mathcal{A} \to \mathbb{Z}_{>0}$, and a subset $\mathfrak{X}$ of intersection points in $\mathcal{A}$ satisfying the following conditions:

(a) $\sum_{H\in\mathcal{A}_i} m_H = \kappa$, independent of i;
(b) for each $H \in \mathcal{A}_i$ and $H' \in \mathcal{A}_j$ with $i \neq j$, $H \cap H' \in \mathfrak{X}$;
(c) for each $x \in \mathfrak{X}$, the sum $\sum_{H\in\mathcal{A}_i, x\in H} m_H$ is independent of i;
(d) for every $1 \leq i \leq k$ and $H, H' \in \mathcal{A}_i$, there is a sequence $\{H_0, H_1, \cdots, H_s\}$ in $\mathcal{A}_i$ with $H_0 = H$ and $H_s = H'$ such that $H_{j-1} \cap H_j \notin \mathfrak{X}$ for all $1 \leq j \leq s$.
(e) $\gcd(m_H \colon H \in \mathcal{A}){=}1$.

Such a multinet is called a (k, κ)-multinet.

Note that the multinet only depends on the intersection lattice of $\mathcal{A}$. By results of Pereira and Yuzvinsky [**PY08**] and Yuzvinsky [**Yuz09**], $k = 3$ or 4 if $|\mathfrak{X}| > 1$. Moreover, Yuzvinsky conjectured that $k = 4$ only happens for Hessian arrangement of 12 planes in $\mathbb{CP}^2$, defined by the polynomial

$$xyz \prod_{j=0}^{2} \prod_{k=0}^{2} (x + \omega^j y + \omega^k z) = 0 \text{ with } \omega = \exp 2\pi i/3.$$

As shown by Falk and Yuzvinsky [**FY07**] and Pereira and Yuzvinsky [**PY08**], every (k, κ)-multinet gives an orbifold map $f \colon X \to C_{0,k} \subset \mathbb{CP}^1$ with $X = \mathbb{CP}^2 - \mathcal{A}$ by

$$f(x) := [\prod_{H\in\mathcal{A}_1} f_H^{m_H}(x), \prod_{H\in\mathcal{A}_2} f_H^{m_H}(x)].$$

DEFINITION 4.6. Let $X = \mathbb{CP}^2 - \mathcal{A}$ with $\mathcal{A}$ being an essential line arrangement in $\mathbb{CP}^2$. Given a hyperbolic orbifold map $f \colon X \to C_{0,k}$,

- we say that f supports $\mathcal{A}$, if it is not the restriction of another orbifold map over the complement of a sub-arrangement of $\mathcal{A}$;
- we say that f is local, if it is the restriction of another hyperbolic orbifold map over the complement of $\mathcal{A}_x$, where x is an intersection point in $\mathcal{A}$ with at least 3 lines passing trough it and $\mathcal{A}_x$ is the arrangement consisting all lines in $\mathcal{A}$ contains x;
- we say that f is global if f is not local.

Clearly, if f supports $\mathcal{A}$, then f is global. The converse is not true in general.

THEOREM 4.7 ([**FY07**]). *Let $\mathcal{A}$ be a essential hyperplane arrangement in $\mathbb{CP}^2$ with $X = \mathbb{CP}^2 - \mathcal{A}$. Then every positive dimensional component of $\mathcal{V}_1^1(X, \mathbb{C})$ passing through the origin is obtained as a pull-back as in Theorem 3.7(a) using either a local hyperbolic orbifold map or a global hyperbolic map $f \colon X \to C_{0,k}$ coming from a multinet structure on a subarrangement of $\mathcal{A}$. Moreover, given a global hyperbolic orbifold map $f \colon X \to C_{0,k}$, either $k = 2$ and f has multiple fibers or $k = 3, 4$. In particular, f supports $\mathcal{A}$ and $k = 3, 4$ if and only if f is obtained from the multinet structure on $\mathcal{A}$.*

The above theorem shows that the existence of global hyperbolic orbifold map $f \colon X \to C_{0,k}$ with $k = 3, 4$ are combinatorially determined (by the multinet structure for subarrangement of $\mathcal{A}$).

QUESTION 4.8. We have the following questions for $X = \mathbb{CP}^2 \setminus \mathcal{A}$ with $\mathcal{A}$ being a essential line arrangement in $\mathbb{CP}^2$.

(i) Is the existence of global hyperbolic orbifold map $f \colon X \to C_{0,2} = \mathbb{C}^*$ combinatorially determined?

(ii) Is the existence of global hyperbolic orbifold map $f\colon X \to C_{0,k}$ for $k = 3, 4$ having multiple fibers combinatorially determined?

Assume that $\mathcal{A}$ has $(d+1)$-many lines and let f_i be the degree one defining equation for the ith line $H_i \in \mathcal{A}$ with $0 \leq i \leq d$.

ASSUMPTION 4.9. For nonzero $a = (a_0, \cdots, a_d) \in \mathbb{Z}^{d+1}$, we assume that $\sum_{i=0}^{d} a_i = 0$ and $\gcd(a_0, \cdots, a_d) = 1$.

Then each $a \in \mathbb{Z}^{d+1}$ satisfying this assumption gives a well-defined algebraic map

$$f^a = [\prod_{a_i>0} f_i^{a_i}, \prod_{a_i<0} f_i^{-a_i}]\colon X \to \mathbb{C}^*.$$

Note that f^a has connected generic fiber, since $\gcd(a_0, \cdots, a_d) = 1$. If we require f^a to be a global hyperbolic orbifold map, then f^a has to be surjective and have multiple fiber. Moreover, every global hyperbolic orbifold map $f\colon X \to C_{0,2} = \mathbb{C}^*$ has to be realized in this way. Note that f^a induces a surjective map

$$f^a_*\colon \pi_1(X) \to \pi_1(\mathbb{C}^*) \cong \mathbb{Z}.$$

This map only depends on a, which sends the meridian of H_i to a_i. It gives a cohomology class in $H^1(X, \mathbb{Z})$, say, denoted by $\nu^a_{\mathbb{Z}}$. Proposition 4.3 shows that f^a has multiple fibers only if $\tau_1(X, \nu^a_{\mathbb{Z}}) > 1$, which leads to the following combinatorics question related to Question 4.8(i).

QUESTION 4.10. For which $a \in \mathbb{Z}^{d+1}$ satisfying Assumption 4.9 do we have $\tau_1(X, \nu^a_{\mathbb{Z}}) > 1$?

On the other hand, we can also use Proposition 4.3 to study Question 4.8(ii). Let $f\colon X \to C_{0,k}$ be a global orbifold map with $k = 3, 4$ and f supports $\mathcal{A}$. In particular, f comes from a multinet on $\mathcal{A}$. Then f induces a surjective map

$$f_*\colon H_1(X, \mathbb{Z}) \to H_1(C_{0,k}, \mathbb{Z}) \cong \mathbb{Z}^{k-1}.$$

Choosing any epimorphism $\nu\colon \mathbb{Z}^{k-1} \to \mathbb{Z}$, we obtain a composed epimorphism, still denoted by ν. Denote corresponding cohomology class in $H^1(X, \mathbb{Z})$ by $\nu_{\mathbb{Z}}$. Proposition 4.3 shows that f has multiple fibers only if $\tau_1(X, \nu_{\mathbb{Z}}) > 1$ for any such ν. Since $k = 3$ or 4, there are infinitely many choices for ν, which gives a great possibility to show that f has no multiple fibers in this case. We confirm this guess under some assumptions for multinet when $k = 3$.

ASSUMPTION 4.11. Given a $(3, \kappa)$-multinet on $\mathcal{A}$ in $\mathbb{CP}^2$, we assume that

(a) $m_H = 1$ for any $H \in \mathcal{A}_1$ or $\mathcal{A}_2$.
(b) If $H, H' \in \mathcal{A}_1$ and $H \cap H' \notin \mathfrak{X}$, then there is no third hyperplane in $\mathcal{A}$ passing through $H \cap H'$. So is $\mathcal{A}_2$.
(c) Consider $\mathcal{A}_3$ and $\mathfrak{X}$. If there exists a point $x \in \mathfrak{X}$ such that x is contained in precisely one hyperplane $H \in \mathcal{A}_3$, one replaces $\mathcal{A}_3$ and $\mathfrak{X}$ by deleting the hyperplane H and the point x, respectively. We assume that one can run this procedure for $\mathcal{A}_3$ and $\mathfrak{X}$ until there is no hyperplane left in $\mathcal{A}_3$.

Under this assumption, the author, jointly with Li, showed that the corresponding orbifold map f has no multiple fibers. The technical conditions in Assumption 4.11 are used to prove $\tau_1(X, \nu_{\mathbb{Z}}) = 1$ for some ν.

PROPOSITION 4.12 ([**LL24**, Proposition 5.9]). *Let $f\colon X \to C_{0,3}$ be an orbifold map corresponding to a $(3,\kappa)$-multinet structure for $\mathcal{A}$ in $\mathbb{CP}^2$ with $X = \mathbb{CP}^2 \setminus \mathcal{A}$. If this multinet satisfies Assumption 4.11, f has no multiple fibers.*

EXAMPLE 4.13. We give a list of examples which satisfies Assumption 4.11: Figure 1(a), Figure 2, and Figure 3(a),(b) in [**FY07**] with A, B, C corresponding to $\mathcal{A}_1, \mathcal{A}_2, \mathcal{A}_3$ and Figure 1 in [**Suc14**]. It is also easy to find examples which do not satisfy Assumption 4.11; see Figure 1(b) in [**FY07**] or the monomial (alias Ceva) arrangement (with $m \geq 3$) in [**Dim17**, Example 6.11].

Consider a global hyperbolic orbifold map $f\colon X \to C_{0,k}$ with $k = 2, 3, 4$. When $k = 2$, there are plenty of examples where f has multiple fibers; see Example 4.4. In fact, Denham and Suciu discovered a combinatorial construction to produce such orbifold maps with $k = 2$ having multiple fibers from a given multinet, called the pointed multinet structure; see, e.g., [**Suc14**, Section 3.6]. On the other hand, the author and Xie discovered that the double star arrangement has a global hyperbolic orbifold map f with $k = 2$ having multiple fibers does not come from the pointed multinet structure [**LX24**]. Note that in all these examples, f does not support $\mathcal{A}$. Moreover, to the author's knowledge, no examples with f having multiple fibers are known when $k = 3, 4$.

QUESTION 4.14. Is it possible to show that f always has no multiple fibers with k being 3 or 4, at least when f supports $\mathcal{A}$?

Proposition 4.12 gives some evidence to this question when f supports $\mathcal{A}$. All the questions in this section should be considered as sub-questions of the following aforementioned question.

QUESTION 4.15. Is $\mathcal{V}_1^1(X, \mathbb{C})$ combinatorially determined for hyperplane arrangement complement X?

5. Proof of results

5.1. Proof of Proposition 2.7. Set $\mathbb{Z}[\mathbf{t}^{\pm}] = \mathbb{Z}[\mathbb{Z}^n]$. Let M be a finitely generated $\mathbb{Z}[\mathbf{t}^{\pm}]$-module. Consider a presentation of M by an exact sequence

$$\mathbb{Z}[\mathbf{t}^{\pm}]^{b'} \xrightarrow{\partial} \mathbb{Z}[\mathbf{t}^{\pm}]^{b} \to M \to 0$$

where ∂ is given by a matrix of size $b' \times b$ with entries in $\mathbb{Z}[\mathbf{t}^{\pm}]$. For any $j \geq 0$, let $\Delta^j(M)$ denote the the greatest common divisor of all the $(b-j)\times(b-j)$ minors of ∂ with the convention that $\Delta^j(M) = 1$ if $j \geq b$. Clearly $\Delta^j(M)$ divides $\Delta^{j-1}(M)$ for every $j \geq 1$. It is known that $\Delta^j(M)$ only depends on M, not on any particular presentation. The rank of the module M over $\mathbb{Z}[\mathbf{t}^{\pm}]$ is the dimension of the vector space $M \otimes_{\mathbb{Z}[\mathbf{t}^{\pm}]} \mathcal{F}'$ over the fractional field $\mathcal{F}'$ of $\mathbb{Z}[\mathbf{t}^{\pm}]$. If M has rank r, then $\Delta^j(M) = 0$ if $j < r$, and $\Delta^j(M) = \Delta^{j-r}(\mathrm{Tor}(M))$, where

$$\mathrm{Tor}(M) = \{x \in M \mid fx = 0 \text{ for some non-zero } f \in \mathbb{Z}[\mathbf{t}^{\pm}]\}.$$

As in [**Le14**, section 1.3], a $\mathbb{Z}[\mathbf{t}^{\pm}]$-module M is called pseudo-zero if for every prime ideal P of height 1, the localization M_P is 0. A $\mathbb{Z}[\mathbf{t}^{\pm}]$-module morphism $M_1 \to M_2$ is a pseudo-isomorphism if the kernel and co-kernel are pseudo-zero. Moreover, two finitely generated torsion $\mathbb{Z}[\mathbf{t}^{\pm}]$-modules M_1 and M_2 are pseudo-isomorphic if and only if $\Delta^j(M_1) = \Delta^j(M_2)$ for any $j \geq 0$.

To prove Proposition 2.7, let us first recall a theorem proved by Le.

THEOREM 5.1 ([**Le14**, Theorem 6]). *Assume that M_1 and M_2 are two pseudo-isomorphic finitely generated $\mathbb{Z}[\mathbf{t}^{\pm}]$ modules. Then $|\mathrm{tor}_{\mathbb{Z}}(M_1 \otimes_{\mathbb{Z}[\mathbf{t}^{\pm}]} \mathbb{Z}[\mathbb{Z}^n/\Gamma])|$ and $|\mathrm{tor}_{\mathbb{Z}}(M_2 \otimes_{\mathbb{Z}[\mathbf{t}^{\pm}]} \mathbb{Z}[\mathbb{Z}^n/\Gamma])|$ have the same growth rate in the sense that*

$$\lim_{\langle\Gamma\rangle\to\infty} \Big(\frac{\log |\mathrm{tor}_{\mathbb{Z}}(M_1 \otimes_{\mathbb{Z}[\mathbf{t}^{\pm}]} \mathbb{Z}[\mathbb{Z}^n/\Gamma])|}{|\mathbb{Z}^n/\Gamma|} - \frac{\log |\mathrm{tor}_{\mathbb{Z}}(M_2 \otimes_{\mathbb{Z}[\mathbf{t}^{\pm}]} \mathbb{Z}[\mathbb{Z}^n/\Gamma])|}{|\mathbb{Z}^n/\Gamma|}\Big) = 0.$$

Here $|\mathrm{tor}_{\mathbb{Z}}(M_i \otimes_{\mathbb{Z}[\mathbf{t}^{\pm}]} \mathbb{Z}[\mathbb{Z}^n/\Gamma])|$ denotes the order of the torsion part of $M_i \otimes_{\mathbb{Z}[t^{\pm}]} \mathbb{Z}[\mathbb{Z}^n/\Gamma]$ as an abelian group.

Note that the limit in this theorem is the ordinary limit, not the upper limit. The following result is a direct application of this theorem and Theorem 2.5.

PROPOSITION 5.2. *Let M be a finitely generated $\mathbb{Z}[\mathbf{t}^{\pm}]$-modules with rank r. If $\Delta^r(M)$ is a product of some generalized cyclotomic polynomials with a nonzero integer c, then*

$$\lim_{\langle\Gamma\rangle\to\infty} \frac{\log |\mathrm{tor}_{\mathbb{Z}}(M \otimes_{\mathbb{Z}[\mathbf{t}^{\pm}]} \mathbb{Z}[\mathbb{Z}^n/\Gamma])|}{|\mathbb{Z}^n/\Gamma|} = \log |c|.$$

PROOF. Consider the submodule $\mathrm{Tor}(M)$ of M. We have that

$$\Delta^j(M) = \Delta^{j-r}(\mathrm{Tor}(M)).$$

Moreover, it follows from Theorem 2.5 and [**Le14**, Theorem 1.2, Proposition 3.4] that

$$\lim_{\langle\Gamma\rangle\to\infty} \Big(\frac{\log |\mathrm{tor}_{\mathbb{Z}}(M \otimes_{\mathbb{Z}[\mathbf{t}^{\pm}]} \mathbb{Z}[\mathbb{Z}^n/\Gamma])|}{|\mathbb{Z}^n/\Gamma|} - \frac{\log |\mathrm{tor}_{\mathbb{Z}}(\mathrm{Tor}(M) \otimes_{\mathbb{Z}[\mathbf{t}^{\pm}]} \mathbb{Z}[\mathbb{Z}^n/\Gamma])|}{|\mathbb{Z}^n/\Gamma|}\Big) = 0.$$

So without loss of generality, we assume that M is a finitely generated torsion $\mathbb{Z}[\mathbf{t}^{\pm}]$-module. Assume that $\Delta^j(M) = 1$ for $j > k$. Consider a finitely generated torsion $\mathbb{Z}[\mathbf{t}^{\pm}]$-module

$$M' := \bigoplus_{j=0}^{k} \frac{\mathbb{Z}[\mathbf{t}^{\pm}]}{(\Delta^j(M)/\Delta^{j+1}(M))}.$$

It is easy to see that $\Delta^j(M) = \Delta^j(M')$ for any $j \geq 0$. Then M is pseudo-isomorphic to M' by [**Hil02**, Theorem 3.5]. Using Theorem 5.1, the proof is reduced to M'. Note that $\Delta^{j+1}(M)$ divides $\Delta^j(M)$ for all $j \geq 0$, hence $\Delta^j(M)/\Delta^{j+1}(M)$ is a also product of some generalized cyclotomic polynomials and some nonzero integer. Since M' is a direct sum, the proof is further reduced to the module with the type $N = \mathbb{Z}[\mathbf{t}^{\pm}]/(h)$, where h a product of some generalized cyclotomic polynomials and some nonzero integer d.

Set $h' = \frac{h}{d}$ and $N' = \mathbb{Z}[\mathbf{t}^{\pm}]/(h')$. Clearly we have

$$|\mathrm{tor}_{\mathbb{Z}}(N \otimes_{\mathbb{Z}[\mathbf{t}^{\pm}]} \mathbb{Z}[\mathbb{Z}^n/\Gamma])| = |d|^{|\mathbb{Z}^n/\Gamma|} \cdot |\mathrm{tor}_{\mathbb{Z}}(N' \otimes_{\mathbb{Z}[\mathbf{t}^{\pm}]} \mathbb{Z}[\mathbb{Z}^n/\Gamma])|.$$

Since h' is a product of some generalized cyclotomic polynomials, using Theorem 2.5 and [**Le14**, Theorem 3] we have

$$\lim_{\langle\Gamma\rangle\to\infty} \frac{\log |\mathrm{tor}_{\mathbb{Z}}(N' \otimes_{\mathbb{Z}[\mathbf{t}^{\pm}]} \mathbb{Z}[\mathbb{Z}^n/\Gamma])|}{|\mathbb{Z}^n/\Gamma|} = 0.$$

Then the claim follows. □

PROOF OF PROPOSITION 2.7. As shown in the proof of [**Le14**, Theorem 4], we have

$$|H_i(X^{\nu,\Gamma},\mathbb{Z})_{\mathrm{tor}}| = |\mathrm{tor}_{\mathbb{Z}}\big(M\otimes_{\mathbb{Z}[t^{\pm}]}\mathbb{Z}[\mathbb{Z}^n/\Gamma]\big)|$$

with $M = \mathrm{coker}\partial_i$, see [**Le14**, (54) on p. 745]. Clearly M has a presentation by the exact sequence

$$C_{i+1}(X^\nu,\mathbb{Z}) \overset{\partial_i}{\to} C_i(X^\nu,\mathbb{Z}) \to M \to 0$$

and $\mathrm{rank}M = \mathrm{rank}C_i(X^\nu,\mathbb{Z}) - \mathrm{rank}\partial_i$. Then we have $\Delta^{\mathrm{rank}M}(M) = \Delta_i(X^\nu)$ by definition. Therefore, the claim follows from Proposition 5.2. □

5.2. Proof of Proposition 2.9. The first equality follows from [**LL23**, Proof of Propsition 2.2]. The second equality follows directly from the fact that computing $\mathrm{rank}H_i(X^\nu,\mathbb{K})$ over the fraction field is same as computing the corresponding homology over a general point.

The second claim follows from the fact that every irreducible component of the zero locus of $\Delta_i(X^\nu)$ in $(\mathbb{C}^*)^n$ gives a prime ideal with height one in $\mathbb{C}[\mathbf{t}^{\pm}]$. Taking localization of $\mathbb{C}[\mathbf{t}^{\pm}]$ over this primed ideal, one gets a PID, say, denoted by R. Consider the localized complex $C_*(X^\nu,\mathbb{C})\otimes_{\mathbb{C}[\mathbf{t}^{\pm}]} R$. Then the second claim follows easily by analysing this localized complex of modules over PID; see, e.g., [**BLW18**, Section 2.2].

5.3. Proof of Proposition 3.3. Let $\mathrm{Pic}^0(X)$ denote the identity component of the Picard variety of X, which parameterize topologically trivial line bundles on X. Set

$$S_k^{p,q}(X) := \{\mathcal{L}\in \mathrm{Pic}^0(X) \mid \dim H^q(X,\Omega_X^p\otimes\mathcal{L})\geq k\}.$$

Green and Lazarsfeld [**GL91**] showed that $S_k^{p,q}(X)$ is a finite union of torsion translated sub-abelian variety of $\mathrm{Pic}^0(X)$. In particular, these are subvarieties of $\mathrm{Pic}^0(X)$.

Consider the moduli space of unitary rank one $\mathbb{C}$-local system on X defined as

$$\mathrm{Uchar}^0(X) := \mathrm{Hom}(H_1(X,\mathbb{Z})/\{\mathrm{torsion}\}, U(1)).$$

Any unitary local system $L\in \mathrm{Uchar}^0(X)$ corresponds to a topologically trivial line bundle $\mathcal{L} = L\otimes_{\mathbb{C}}\mathcal{O}_X$. This gives a one-to-one correspondence between $\mathrm{Uchar}^0(X)$ and $\mathrm{Pic}^0(X)$ as real torus. By slight modification of the usual Hodge theoretic arguments, one gets the following Hodge decomposition

$$H^i(X,L)\cong \bigoplus_{p+q=i} H^q(X,\Omega_X^p\otimes\mathcal{L}). \tag{5.1}$$

The epimorphism ν induces an embedding $\nu^\#\colon (U(1))^n \hookrightarrow \mathrm{Uchar}^0(X)$. Let $\mathrm{im}\nu^\#$ denote the corresponding image in $\mathrm{Pic}^0(X)$ by the above identification. Since $S_k^{p,q}(X)$ are subvarieties of $\mathrm{Pic}^0(X)$, we can pick a generic unitary local system $L'\in\mathrm{im}\nu^\#$ with $\mathcal{L}' = L'\otimes\mathcal{O}_X$ (to be precise, L' is generic with respect to all degrees (p,q) with $0\leq p,q\leq \dim X$). By semicontinuity, for any unitary local system L and its corresponding line bundle $\mathcal{L}$ we have

$$\dim H^q(X,\Omega_X^p\otimes\mathcal{L}')\leq \dim H^q(X,\Omega_X^p\otimes\mathcal{L}).$$

Let $\pi\colon X^{\nu,\Gamma}\to X$ denote the covering map. Then

$$\pi_*\mathcal{O}_X = \bigoplus \mathcal{L}_\lambda$$

where $\mathcal{L}_\lambda = L_\lambda \otimes \mathcal{O}_X$ and the direct sum runs over all unitary local system L_λ with $\lambda \in \mathrm{Hom}(\mathbb{Z}^n/\Gamma, \mathbb{C}^*)$. We have

$$\dim H^q(X^\Gamma, \Omega^p_{X^\Gamma}) = \sum_\lambda \dim H^q(X, \Omega^p_X \otimes \mathcal{L}_\lambda) \geq |\mathbb{Z}^n/\Gamma| \cdot \dim H^q(X, \Omega^p_X \otimes \mathcal{L}').$$

Meanwhile, the Hodge decomposition gives us

$$\sum_{p+q=i} \dim H^q(X^\Gamma, \Omega^p_{X^\Gamma}) = \dim H^i(X^\Gamma, \mathbb{C}).$$

Putting all together, we have that

$$\sum_{p+q=i} \dim H^q(X, \Omega^p_X \otimes \mathcal{L}') \leq \sum_{p+q=i} \frac{\dim H^q(X^\Gamma, \Omega^p_{X^\Gamma})}{|\mathbb{Z}^n/\Gamma|} = \frac{\dim H^i(X^\Gamma, \mathbb{C})}{|\mathbb{Z}^n/\Gamma|}.$$

$\mathcal{V}^i_k(X)$ are also closed subvarieties. By further requiring the unitary local system L' generic with respect to the jump loci, Proposition 2.9 and the universal coefficients theorem imply that the right hand of the above inequality converges to $\dim H^i(X, L')$ when $\langle\Gamma\rangle \to \infty$. Note that the Hodge decomposition (5.1) gives us

$$\sum_{p+q=i} \dim H^q(X, \Omega^p_X \otimes \mathcal{L}') = \dim H^i(X, L').$$

Therefore we have

$$\lim_{\langle\Gamma\rangle\to\infty} \frac{h^{p,q}(X^{\nu,\Gamma})}{|\mathbb{Z}^n/\Gamma|} = \dim H^q(X, \Omega^p_X \otimes \mathcal{L}').$$

References

[Ara97] D. Arapura, *Geometry of cohomology support loci for local systems. I*, J. Algebraic Geom. **6** (1997), no. 3, 563–597. MR1487227

[AW23] D. Arapura and B. Wang, *Perverse sheaves on varieties with large fundamental groups*, To appear in Journal of Diff. Geom.

[AMW23] D. Arapura, L. Maxim, and B. Wang, *Hodge-theoretic variants of the Hopf and Singer Conjectures*. arXiv:2310.14131.

[ACM10] E. Artal Bartolo and J. I. Cogolludo-Agustín, *On the connection between fundamental groups and pencils with multiple fibers*, J. Singul. **2** (2010), 1–18, DOI 10.5427/jsing.2010.2a. MR2763015

[ACM13] E. Artal Bartolo, J. I. Cogolludo-Agustín, and D. Matei, *Characteristic varieties of quasi-projective manifolds and orbifolds*, Geom. Topol. **17** (2013), no. 1, 273–309, DOI 10.2140/gt.2013.17.273. MR3035328

[Ati76] M. F. Atiyah, *Elliptic operators, discrete groups and von Neumann algebras*, Colloque "Analyse et Topologie" en l'Honneur de Henri Cartan (Orsay, 1974), Astérisque, No. 32-33, Soc. Math. France, Paris, 1976, pp. 43–72. MR420729

[Bea92] A. Beauville, *Annulation du H^1 pour les fibrés en droites plats* (French), Complex algebraic varieties (Bayreuth, 1990), Lecture Notes in Math., vol. 1507, Springer, Berlin, 1992, pp. 1–15, DOI 10.1007/BFb0094507. MR1178716

[BNS87] R. Bieri, W. D. Neumann, and R. Strebel, *A geometric invariant of discrete groups*, Invent. Math. **90** (1987), no. 3, 451–477, DOI 10.1007/BF01389175. MR914846

[Bud09] N. Budur, *Unitary local systems, multiplier ideals, and polynomial periodicity of Hodge numbers*, Adv. Math. **221** (2009), no. 1, 217–250, DOI 10.1016/j.aim.2008.12.006. MR2509325

[BLW18] N. Budur, Y. Liu, and B. Wang, *The monodromy theorem for compact Kähler manifolds and smooth quasi-projective varieties*, Math. Ann. **371** (2018), no. 3-4, 1069–1086, DOI 10.1007/s00208-017-1541-3. MR3831264

[BW15] N. Budur and B. Wang, *Cohomology jump loci of quasi-projective varieties* (English, with English and French summaries), Ann. Sci. Éc. Norm. Supér. (4) **48** (2015), no. 1, 227–236, DOI 10.24033/asens.2242. MR3335842

[BW17] N. Budur and B. Wang, *Recent results on cohomology jump loci*, Hodge theory and L^2-analysis, Adv. Lect. Math. (ALM), vol. 39, Int. Press, Somerville, MA, 2017, pp. 207–243. MR3751292

[BW20] N. Budur and B. Wang, *Absolute sets and the decomposition theorem* (English, with English and French summaries), Ann. Sci. Éc. Norm. Supér. (4) **53** (2020), no. 2, 469–536, DOI 10.24033/asens.2426. MR4094563

[CO00] D. C. Cohen and P. Orlik, *Arrangements and local systems*, Math. Res. Lett. **7** (2000), no. 2-3, 299–316, DOI 10.4310/MRL.2000.v7.n3.a5. MR1764324

[CS99] D. C. Cohen and A. I. Suciu, *Characteristic varieties of arrangements*, Math. Proc. Cambridge Philos. Soc. **127** (1999), no. 1, 33–53, DOI 10.1017/S0305004199003576. MR1692519

[Del08] T. Delzant, *Trees, valuations and the Green-Lazarsfeld set*, Geom. Funct. Anal. **18** (2008), no. 4, 1236–1250, DOI 10.1007/s00039-008-0679-2. MR2465689

[Del10] T. Delzant, *L'invariant de Bieri-Neumann-Strebel des groupes fondamentaux des variétés kählériennes* (French, with French summary), Math. Ann. **348** (2010), no. 1, 119–125, DOI 10.1007/s00208-009-0468-8. MR2657436

[DS14] G. Denham and A. I. Suciu, *Multinets, parallel connections, and Milnor fibrations of arrangements*, Proc. Lond. Math. Soc. (3) **108** (2014), no. 6, 1435–1470, DOI 10.1112/plms/pdt058. MR3218315

[DL23] L. F. Di Cerbo and L. Lombardi *Singer Conjecture for Varieties with Semismall Albanese Map and Residually Finite Fundamental Group*, Proceedings of the Royal Society of Edinburgh: Section A Mathematics. Published online 2024:1-7. doi:10.1017/prm.2024.52

[DL24] L. F. Di Cerbo and L. Lombardi, *L^2-Betti numbers and convergence of normalized Hodge numbers via the weak generic Nakano vanishing theorem* (English, with English and French summaries), Ann. Inst. Fourier (Grenoble) **74** (2024), no. 1, 423–449, DOI 10.5802/aif.3594. MR4748177

[Dim07] A. Dimca, *Characteristic varieties and constructible sheaves*, Atti Accad. Naz. Lincei Rend. Lincei Mat. Appl. **18** (2007), no. 4, 365–389, DOI 10.4171/RLM/503. MR2349994

[Dim09] A. Dimca, *On admissible rank one local systems*, J. Algebra **321** (2009), no. 11, 3145–3157, DOI 10.1016/j.jalgebra.2008.01.039. MR2510044

[Dim17] A. Dimca, *Hyperplane arrangements*, Universitext, Springer, Cham, 2017. An introduction, DOI 10.1007/978-3-319-56221-6. MR3618796

[DP03] A. Dimca and S. Papadima, *Hypersurface complements, Milnor fibers and higher homotopy groups of arrangments*, Ann. of Math. (2) **158** (2003), no. 2, 473–507, DOI 10.4007/annals.2003.158.473. MR2018927

[DP14] A. Dimca and Ş. Papadima, *Non-abelian cohomology jump loci from an analytic viewpoint*, Commun. Contemp. Math. **16** (2014), no. 4, 1350025, 47, DOI 10.1142/S0219199713500259. MR3231055

[DPS09] A. Dimca, Ş. Papadima, and A. I. Suciu, *Topology and geometry of cohomology jump loci*, Duke Math. J. **148** (2009), no. 3, 405–457, DOI 10.1215/00127094-2009-030. MR2527322

[FY07] M. Falk and S. Yuzvinsky, *Multinets, resonance varieties, and pencils of plane curves*, Compos. Math. **143** (2007), no. 4, 1069–1088, DOI 10.1112/S0010437X07002722. MR2339840

[GL91] M. Green and R. Lazarsfeld, *Higher obstructions to deforming cohomology groups of line bundles*, J. Amer. Math. Soc. **4** (1991), no. 1, 87–103, DOI 10.2307/2939255. MR1076513

[Hil02] J. Hillman, *Algebraic invariants of links*, Series on Knots and Everything, vol. 32, World Scientific Publishing Co., Inc., River Edge, NJ, 2002, DOI 10.1142/9789812776648. MR1932169

[Le14] T. Le, *Homology torsion growth and Mahler measure*, Comment. Math. Helv. **89** (2014), no. 3, 719–757, DOI 10.4171/CMH/332. MR3260847

[LL24] F. Li and Y. Liu, *L^2-type invariants and cohomology jump loci for smooth complex quasi-projective varieties*, Preprint, `arXiv:2110.03356`, (2024).

[LL23] F. Li and Y. Liu, *The homology growth for finite abelian covers of smooth quasi-projective varieties*, Preprint, `arXiv:2311.11593`, (2023).

[LX24] Y. Liu and W. Xie, *Double star arrangement and the pointed multinet*, Preprint, arXiv:2409.04032, (2024).

[Lib02] A. Libgober, *First order deformations for rank one local systems with a non-vanishing cohomology*, Topology Appl. **118** (2002), no. 1-2, 159–168, DOI 10.1016/S0166-8641(01)00048-7. Arrangements in Boston: a Conference on Hyperplane Arrangements (1999). MR1877722

[LY00] A. Libgober and S. Yuzvinsky, *Cohomology of the Orlik-Solomon algebras and local systems*, Compositio Math. **121** (2000), no. 3, 337–361, DOI 10.1023/A:1001826010964. MR1761630

[LLS11] P. Linnell, W. Lück, and R. Sauer, *The limit of $\mathbb{F}_p$-Betti numbers of a tower of finite covers with amenable fundamental groups*, Proc. Amer. Math. Soc. **139** (2011), no. 2, 421–434, DOI 10.1090/S0002-9939-2010-10689-5. MR2736326

[LMW21] Y. Liu, L. Maxim, and B. Wang, *Aspherical manifolds, Mellin transformation and a question of Bobadilla-Kollár*, J. Reine Angew. Math. **781** (2021), 1–18, DOI 10.1515/crelle-2021-0055. MR4343101

[Luc94] W. Lück, *Approximating L^2-invariants by their finite-dimensional analogues*, Geom. Funct. Anal. **4** (1994), no. 4, 455–481, DOI 10.1007/BF01896404. MR1280122

[Luc02] W. Lück, *L^2-invariants: theory and applications to geometry and K-theory*, Ergebnisse der Mathematik und ihrer Grenzgebiete. 3. Folge. A Series of Modern Surveys in Mathematics [Results in Mathematics and Related Areas. 3rd Series. A Series of Modern Surveys in Mathematics], vol. 44, Springer-Verlag, Berlin, 2002, DOI 10.1007/978-3-662-04687-6. MR1926649

[Luc13] W. Lück, *Approximating L^2-invariants and homology growth*, Geom. Funct. Anal. **23** (2013), no. 2, 622–663, DOI 10.1007/s00039-013-0218-7. MR3053758

[Luc16] W. Lück, *Approximating L^2-invariants by their classical counterparts*, EMS Surv. Math. Sci. **3** (2016), no. 2, 269–344, DOI 10.4171/EMSS/18. MR3576534

[Mah60] K. Mahler, *An application of Jensen's formula to polynomials*, Mathematika **7** (1960), 98–100, DOI 10.1112/S0025579300001637. MR124467

[Mah62] K. Mahler, *On some inequalities for polynomials in several variables*, J. London Math. Soc. **37** (1962), 341–344, DOI 10.1112/jlms/s1-37.1.341. MR138593

[OS80] P. Orlik and L. Solomon, *Combinatorics and topology of complements of hyperplanes*, Invent. Math. **56** (1980), no. 2, 167–189, DOI 10.1007/BF01392549. MR558866

[PS10] S. Papadima and A. I. Suciu, *Bieri-Neumann-Strebel-Renz invariants and homology jumping loci*, Proc. Lond. Math. Soc. (3) **100** (2010), no. 3, 795–834, DOI 10.1112/plms/pdp045. MR2640291

[PY08] J. V. Pereira and S. Yuzvinsky, *Completely reducible hypersurfaces in a pencil*, Adv. Math. **219** (2008), no. 2, 672–688, DOI 10.1016/j.aim.2008.05.014. MR2435653

[PS13] M. Popa and C. Schnell, *Generic vanishing theory via mixed Hodge modules*, Forum Math. Sigma **1** (2013), Paper No. e1, 60, DOI 10.1017/fms.2013.1. MR3090229

[Rai12] J. Raimbault, *Exponential growth of torsion in abelian coverings*, Algebr. Geom. Topol. **12** (2012), no. 3, 1331–1372, DOI 10.2140/agt.2012.12.1331. MR2966689

[Ran02] R. Randell, *Morse theory, Milnor fibers and minimality of hyperplane arrangements*, Proc. Amer. Math. Soc. **130** (2002), no. 9, 2737–2743, DOI 10.1090/S0002-9939-02-06412-2. MR1900880

[Sch95] K. Schmidt, *Dynamical systems of algebraic origin*, Progress in Mathematics, vol. 128, Birkhäuser Verlag, Basel, 1995. MR1345152

[Sch15] C. Schnell, *Holonomic D-modules on abelian varieties*, Publ. Math. Inst. Hautes Études Sci. **121** (2015), 1–55, DOI 10.1007/s10240-014-0061-x. MR3349829

[Sim93A] C. Simpson, *Subspaces of moduli spaces of rank one local systems*, Ann. Sci. École Norm. Sup. (4) **26** (1993), no. 3, 361–401. MR1222278

[Sim93B] C. Simpson, *Lefschetz theorems for the integral leaves of a holomorphic one-form*, Compositio Math. **87** (1993), no. 1, 99–113. MR1219454

[Suc01] A. I. Suciu, *Fundamental groups of line arrangements: enumerative aspects*, Advances in algebraic geometry motivated by physics (Lowell, MA, 2000), Contemp. Math., vol. 276, Amer. Math. Soc., Providence, RI, 2001, pp. 43–79, DOI 10.1090/conm/276/04510. MR1837109

[Suc02] A. I. Suciu, *Translated tori in the characteristic varieties of complex hyperplane arrangements*, Topology Appl. **118** (2002), no. 1-2, 209–223, DOI 10.1016/S0166-8641(01)00052-9. Arrangements in Boston: a Conference on Hyperplane Arrangements (1999). MR1877726

[Suc11] A. I. Suciu, *Fundamental groups, Alexander invariants, and cohomology jumping loci*, Topology of algebraic varieties and singularities, Contemp. Math., vol. 538, Amer. Math. Soc., Providence, RI, 2011, pp. 179–223, DOI 10.1090/conm/538/10600. MR2777821

[Suc14] A. I. Suciu, *Hyperplane arrangements and Milnor fibrations* (English, with English and French summaries), Ann. Fac. Sci. Toulouse Math. (6) **23** (2014), no. 2, 417–481, DOI 10.5802/afst.1412. MR3205599

[Suc21] A. I. Suciu, *Sigma-invariants and tropical varieties*, Math. Ann. **380** (2021), no. 3-4, 1427–1463, DOI 10.1007/s00208-021-02172-z. MR4297190

[Yuz09] S. Yuzvinsky, *A new bound on the number of special fibers in a pencil of curves*, Proc. Amer. Math. Soc. **137** (2009), no. 5, 1641–1648, DOI 10.1090/S0002-9939-08-09753-0. MR2470822

Institute of Geometry and Physics, University of Science and Technology of China, 96 Jinzhai Road, Hefei Anhui 230026, People's Republic of China

Email address: `liuyq@ustc.edu.cn`

Research Articles

Contemporary Mathematics
Volume **816**, 2025
https://doi.org/10.1090/conm/816/16353

Hodge-theoretic variants of the Hopf and Singer conjectures

Donu Arapura, Laurenţiu G. Maxim, and Botong Wang

ABSTRACT. We propose several Hodge-theoretic analogs of the conjectures of Hopf and Singer, and prove them in some special cases.

1. Introduction

1.1. Singer–Hopf conjecture. In 1931, Hopf made the following conjecture on the sign of the Euler characteristic of a Riemannian manifold, formulated here as strengthened by Singer in the context of aspherical manifolds, e.g., see [**18**, Conjecture 25.1].

CONJECTURE 1.1 (Singer–Hopf). *If X is a closed aspherical manifold of real dimension $2n$, then*

$$(-1)^n \cdot \chi(X) \geq 0. \tag{1.1}$$

Recall that a connected CW complex is said to be aspherical if its universal cover is contractible, so the above conjecture applies more generally to topological manifolds. Conjecture 1.1 is clearly true if $n = 1$. Examples of aspherical manifolds include closed Riemannian manifolds with nonpositive sectional curvature, for which the conjecture is true if $n = 2$ since the Gauss–Bonnet integrand has the desired sign, cf. [**10**, Theorem 5]. However, Conjecture 1.1 is not yet known for all closed aspherical 4-manifolds, and it is open for $n \geq 3$.

Gromov [**16**] introduced the notion of Kähler hyperbolicity, which includes compact Kähler manifolds with negative sectional curvature as a special case, and he verified Conjecture 1.1 for Kähler hyperbolic manifolds. Cao and Xavier [**7**] and Jost and Zuo [**21**] introduced the concept of Kähler nonellipticity, including compact Kähler manifolds with nonpositive sectional curvature, and proved the Singer–Hopf conjecture in this case. See also [**13**]. These works proved the desired sign of the Euler characteristic by means of vanishing L^2-cohomology. More recently, Liu, Maxim, and Wang [**28**] proved the complex projective version of Conjecture 1.1

2020 *Mathematics Subject Classification.* Primary 14F35, 14F45, 14C30, 14J45, 32S60.

Key words and phrases. Singer–Hopf conjecture, aspherical manifold, Euler characteristic, sectional curvature, nef bundle, Fano variety.

The first author was partially supported by a grant from the Simons Foundation.

The second author was supported by the project “Singularities and Applications” - CF 132/31.07.2023 funded by the European Union - NextGenerationEU - through Romania’s National Recovery and Resilience Plan.

under the assumption that the (holomorphic) cotangent bundle T^*X of X is numerically effective (nef), e.g., globally generated. Moreover, it was conjectured in [**28**, Section 6] that aspherical complex projective manifolds have nef cotangent bundles. Building on ideas of [**12**] and [**28**], Arapura and Wang [**2**] gave a new proof of Conjecture 1.1 for compact Kähler manifolds with nonpositive sectional curvature, using the fact that such manifolds have nef cotangent bundles. Several generalizations of the Singer–Hopf conjecture in the singular context have been recently proposed in [**29**, Conjectures 1.2 and 1.3] by using constructible functions.

1.2. Hodge-theoretic variants of the Singer–Hopf conjecture. One of the aims of this note is to formulate Hodge-theoretic analogs of Conjecture 1.1. For this purpose, we only consider compact Kähler (e.g., complex projective) manifolds.

For a compact Kähler manifold X of complex dimension n, the Euler characteristic can be computed as

$$\chi(X) = \sum_{p \geq 0} (-1)^p \cdot \chi^p(X), \tag{1.2}$$

with $\chi^p(X) := \chi(X, \Omega_X^p)$, and $\chi^p(X) = (-1)^n \chi^{n-p}(X)$ by Serre duality. Here, Ω_X^p denotes the sheaf of holomorphic p-forms on X, and note that $\chi^0(X)$ is just the Todd genus of X.

We propose the following natural Hodge-theoretic enhancement of the Singer–Hopf Conjecture 1.1.

CONJECTURE 1.2. *If X is a compact Kähler manifold of dimension n which is aspherical or has a nef cotangent bundle, then for any $0 \leq p \leq n$ one has:*

$$(-1)^{n-p} \cdot \chi^p(X) \geq 0. \tag{1.3}$$

Via the Riemann–Roch theorem, (1.3) yields inequalities for the Chern numbers of X (of weighted degree $2n$), usually referred to as *Arakelov type inequalities.* Moreover, if $p = n$, (1.3) reduces to the nonnegativity of the holomorphic Euler characteristic $\chi(X, K_X) = (-1)^n \cdot \chi(X, \mathcal{O}_X)$ of the canonical bundle K_X, which Kollár [**24**] conjectured for complex projective manifolds with generically large fundamental groups (we refer the reader to [**24**] for the relevant definitions). Recall here that if X is an aspherical projective manifold, then $\pi_1(X)$ is large, e.g., see [**28**, Proposition 6.7]. Let us also mention that it was shown in [**46**, Theorem 4] that if X is projective with Ω_X^1 nef and $\chi(X, K_X) > 0$ then K_X is ample, with the converse statement being conjectured in [**46**] and proved in the special case when the cotangent bundle is globally generated (cf. [**46**, Theorem 3]) or if $\dim X \leq 4$.

Conjecture 1.2 is implied by work of Gromov [**16**] when X is Kähler hyperbolic, and by Jost and Zuo [**21**] in the Kähler nonelliptic case, see also [**13**]. The works [**13, 16, 21**] prove such inequalities via vanishing theorems for L^2-cohomology.

More recently, Popa and Schnell [**37**] proved (1.3) for a complex projective manifold X with a semismall Albanese map, by showing a Nakano-type generic vanishing theorem for the sheaves Ω_X^p of holomorphic p-forms. Note that if the cotangent bundle is globally generated, then the Albanese map of X is an immersion, and (1.3) follows in this case from work of Popa and Schnell [**37**].

In Proposition 2.4 we show that Conjecture 1.2 holds for complex projective curves and surfaces with nef cotangent bundles, but see also Remark 2.5 for the case of aspherical smooth complex compact surfaces. A similar argument can be used to show the nonnegativity of $\chi(X, K_X)$ for X a complex projective manifold with nef cotangent bundle and complex dimension $n \leq 4$ (see Theorem 2.6).

One can further generalize the statement of Conjecture 1.2 as follows.

CONJECTURE 1.3. *If X is a compact Kähler (or complex projective) manifold which is aspherical or has a nef cotangent bundle, and $\mathcal{M}$ is a mixed Hodge module on X, then for any integer p one has:*

$$\chi(X, Gr_F^p DR(\mathcal{M})) \geq 0, \tag{1.4}$$

where $Gr_F^p DR(\mathcal{M})$ are the graded pieces, with respect to the Hodge filtration,[1] *of the de Rham complex associated to $\mathcal{M}$.*

For instance, if $n = \dim X$ and $\mathcal{M} = \mathbb{Q}_X^H[n]$ is the constant Hodge module, then $Gr_F^p DR(\mathcal{M}) \cong \Omega^p[n-p]$, so Conjecture 1.3 reduces in this case to Conjecture 1.2. Work of Popa and Schnell [**37**] on generic vanishing via mixed Hodge modules can be used to prove the inequality (1.4) in the case when X is a smooth subvariety of, or more generally, admits a finite morphism to an abelian variety (see Theorem 2.7).

1.3. Hopf conjecture for positive sectional curvature and generalizations. We also propose a Hodge theoretic variant of another conjecture of Hopf, which can be stated as follows (see, e.g., [**45**]).

CONJECTURE 1.4 (Hopf). *A compact, even-dimensional Riemannian manifold with positive sectional curvature has positive Euler characteristic. A compact, even-dimensional Riemannian manifold with nonnegative sectional curvature has nonnegative Euler characteristic.*

If X is a compact Kähler manifold with nonnegative (resp., positive) sectional curvature, then the holomorphic tangent bundle TX is nef (resp., ample). So the following result from [**12**], which is derived from (semi)positivity properties of nef/ample vector bundles (cf. [**4, 12, 14**]), can be seen as a generalization in the compact Kähler context of Conjecture 1.4.

THEOREM 1.5. *If X is a compact Kähler (resp., complex projective) manifold with a nef (resp., ample) tangent bundle TX (e.g., X has a nonnegative (resp., positive) sectional curvature), then $\chi(X) \geq 0$ (resp., > 0).*

Let us remark that, in the case when X has positive bisectional curvature (e.g., if X has positive sectional curvature), Siu and Yau's proof [**40**] of Frankel's conjecture shows that X is biholomorphic to a complex projective space. The same conclusion was proved by Mori [**32**] for complex projective manifolds with ample tangent bundles. So the positive version of Theorem 1.5 can also be deduced from these deep results. Furthermore, a classification of all compact Kähler manifolds with nonnegative bisectional curvature was obtained by Mok [**31**] (see also [**17**]), and this can be used as in Theorem 2.9 to prove Theorem 1.5 in this situation. The general nef case follows, e.g., from [**12**, Proposition 2.1] and the Gauss–Bonnet theorem.

In this note, we propose the following conjectural statement, which provides a Hodge-theoretic analog of Theorem 1.5, and which holds, e.g., in the case of

[1] Here we use the notation Gr_F^p in place of Gr_{-p}^F corresponding to the identification $F^p = F_{-p}$, which makes the transition between Saito's increasing filtration F_{-p} appearing in the D-module language and the classical Hodge-theoretic situation of a decreasing filtration F^p.

compact Kähler manifolds with nonnegative bisectional curvature (e.g., nonnegative sectional curvature); see Theorem 2.9 for a proof in this case.

CONJECTURE 1.6. *If X is a compact Kähler manifold of dimension n with a nef tangent bundle, then for $0 \leq p \leq n$ one has*

$$(-1)^p \cdot \chi^p(X) \geq 0, \tag{1.5}$$

with $\chi^p(X) := \chi(X, \Omega_X^p)$.

As we discuss in Section 2.3, by results of [**12**] it suffices to prove Conjecture 1.6 in the case when X is a Fano manifold (i.e., a projective algebraic manifold with K_X^{-1} ample). In Proposition 2.8, we prove this remaining case of Conjecture 1.6 under the additional assumption that X has an (algebraic) *cellular decomposition*, in the sense that there is a chain of Zariski closed subsets X_i such that $X_i \setminus X_{i+1}$ is a union of affine spaces (see, e.g., [**15**, Example 1.9.1]). Examples of such varieties include *rational homogeneous* projective manifolds, i.e., projective quotients G/P, with G a semisimple Lie group and $P \subset G$ is a parabolic subgroup (see [**5**, Section 3]). In fact, one has the following conjectural statement of Campana and Peternell [**6**], which, in view of Proposition 2.8, would complete the proof of Conjecture 1.6.

CONJECTURE 1.7 (Campana and Peternell). *Any Fano manifold with nef tangent bundle is rational homogeneous.*

This is a natural generalization to the nef case of the above-mentioned result of Mori [**32**] (formerly known as the Hartshorne conjecture), and it is currently proved up to dimension 5 and in several more cases; see, e.g., [**11, 22, 23, 27, 34, 36, 42, 43**] for recent progress, as well as [**33**] for a survey on this conjecture. Furthermore, in the unpublished preprint [**11**], Conjecture 1.7 is reduced to proving the nonvanishing of the top Segre class of X (cf. [**11**, Theorem 1.2, Conjecture 1.3]).

Given that in order to prove Conjecture 1.6 we do not need the full strength of Conjecture 1.7, we propose here the following weaker statements which imply Conjecture 1.6.

CONJECTURE 1.8. *Any Fano manifold X with nef tangent bundle has an (algebraic) cellular decomposition.*

CONJECTURE 1.9. *If X is a Fano manifold with nef tangent bundle, then its Hodge numbers are concentrated on the diagonal, i.e., $h^{p,q}(X) = 0$ if $p \neq q$.*

For example, Conjecture 1.8 is true if X has nonnegative bisectional curvature. This is an application of Mok's proof of the generalized Frankel conjecture [**31**], using the fact that a Fano manifold is simply connected. This observation already leads to the proof of Conjecture 1.6 when X has nonnegative bisectional curvature, see Theorem 2.9. Moreover, the proof of Proposition 2.8 shows that Conjecture 1.8 implies Conjecture 1.9.

Let us also note that, by results of Carrell and Liebermann [**9**], in order to prove Conjecture 1.6, it would also suffice to show that X admits a holomorphic vector field with only isolated zeros; see Remark 2.10. This latter fact was conjectured by Carrell [**8**] to be equivalent to X being rational (see [**19, 25**] for progress on this conjecture).

We conclude this introduction with yet another conjecture, which may be of independent interest, and which is easily implied by either Conjecture 1.8 or Conjecture 1.9, so in particular it is true whenever any of these conjectures hold (e.g., if X has nonnegative bisectional curvature).

CONJECTURE 1.10. *If X is a Fano manifold with nef tangent bundle, then its odd Betti numbers vanish.*

2. Proofs of results

In this section, we discuss various aspects of the Hodge theoretic variants of the Singer and Hopf conjectures, namely Conjectures 1.2, 1.3 and 1.6 mentioned in the introduction.

We begin by recalling the following definition.

DEFINITION 2.1. If E is a vector bundle on a projective (or compact complex) manifold X, denote by $\mathbf{P}(E)$ the projective bundle of hyperplanes in the fibers of E. A vector bundle E on X is called *ample* (resp., *nef*) if the line bundle $\mathcal{O}(1)$ on $\mathbf{P}(E)$ is ample (resp., nef).

In the complex manifold case, the nefness of a line bundle is considered in the sense of [**12**, Definition 1.2], and this coincides with the usual definition when X is projective. Note that globally generated bundles are nef. Properties of nef bundles are studied, e.g., in [**12**, **26**].

2.1. On Conjecture 1.2.

In this section we restrict our attention to the projective context (but see also Remark 2.5). We will make use of positivity properties for nef vector bundles, which we now recall.

DEFINITION 2.2. The *Schur polynomial* $P_{\underline{a}} \in \mathbb{Z}[c_1, \ldots, c_n]$ of weighted degree $2n$ (with $\deg c_i = 2i$) corresponding to a partition $\underline{a} = (a_1, \ldots, a_n)$ of n, i.e.,

$$n \geq a_1 \geq a_2 \geq \cdots \geq a_n \geq 0, \text{ with } \sum_j a_j = n,$$

is defined as the $n \times n$ determinant

$$P_{\underline{a}}(c) = \det \left(c_{a_i - i + j}\right)_{1 \leq i,j \leq n} .$$

Here, $c_i = 0$ if $i \notin [0, n]$ and $c_0 = 1$.[2]

Then, as a special case of [**12**, Theorem 2.5], one has the following.

THEOREM 2.3. *Let X be a compact Kähler manifold of complex dimension n. Then the Schur polynomials $P_{\underline{a}}(c(E))$ of weighted degree $2n$ in the Chern classes of a nef vector bundle E of rank n on X are nonnegative, that is,*

$$\int_X P_{\underline{a}}(c(E)) \geq 0. \tag{2.1}$$

For simplicity, in what follows, we will drop the integral symbol.

The Riemann–Roch theorem, positivity results from [**12**] as recalled in Theorem 2.3, and the Bogomolov–Miyaoka–Yau inequality[3] [**30**, **41**, **44**] yield the following proposition.

PROPOSITION 2.4. *Conjecture* 1.2 *is true for smooth projective curves and surfaces with nef cotangent bundles.*

---

[2]The Schur polynomial corresponding to the partition $\underline{1}^n = (1, \ldots, 1)$ is the Segre class s_n.

[3]Almost all results concerning the Bogomolov–Miyaoka–Yau inequality are proved under the projective assumption. For a recent extension to compact Kähler manifolds under a suitable positivity condition on the canonical bundle, see [**35**].

PROOF. In the calculations below, we use the notation

$$c_i := c_i(\Omega_X^1) = (-1)^i c_i(TX).$$

Assume $\dim X = 1$. The Riemann–Roch theorem and the nonnegativity result of Theorem 2.3 yield:

$$\chi(X, \Omega_X^1) = -\chi(X, \mathcal{O}_X) = \frac{c_1}{2} \geq 0.$$

Assume $\dim X = 2$. The nonnegativity of Schur polynomials of T^*X implies that $c_2 \geq 0$ and $c_1^2 - c_2 \geq 0$. Together with the Riemann–Roch theorem, this yields:

$$\chi(X, \Omega_X^2) = \chi(X, \mathcal{O}_X) = \frac{c_1^2 + c_2}{12} \geq 0. \tag{2.2}$$

Also, one gets by Riemann–Roch that

$$\chi(X, \Omega_X^1) = \frac{c_1^2 - 5c_2}{6}. \tag{2.3}$$

The nonpositivity of the above expression follows from $c_2 \geq 0$, by using the Bogomolov–Miyaoka–Yau inequality (e.g., in the sense of [**41**] since K_X is nef, hence X is minimal), which in our notations can be written as:

$$c_1^2 \leq 3c_2.$$

□

REMARK 2.5. As pointed out by the referee (cf. also [**1**, Remark 5.6]), Conjecture 1.2 also holds for all aspherical smooth complex compact surfaces. Let us briefly indicate the argument here. Let X be a smooth complex compact surface, and denote by χ and σ its topological Euler characteristic and signature, respectively. Then using (2.2), (2.3), the fact that the Chern numbers c_1^2 and c_2 are the same for both TX and T^*X, together with the standard identities $2\chi + 3\sigma = c_1^2$ and $c_2 = \chi$, one gets that

$$\chi(X, \Omega_X^2) = \chi(X, \mathcal{O}_X) = \frac{1}{4}(\chi + \sigma), \quad \chi(X, \Omega_X^1) = \frac{1}{2}(\sigma - \chi). \tag{2.4}$$

Moreover, one has by [**20**, Theorem 2] that $\chi \geq |\sigma|$, unless X is a ruled surface over a curve of genus ≥ 2. Since ruled surfaces are not aspherical, one then concludes that Conjecture 1.2 holds for any aspherical smooth complex compact surface X. Finally, as in [**1**, Remark 5.6], one further observes that if X is an aspherical smooth compact complex surface of general type then the inequalities in Conjecture 1.2 become strict.

We next show that the arguments of Proposition 2.4 can be extended to dimensions $n = 3$ and $n = 4$, to show the nonnegativity of $\chi(X, K_X)$ in these dimensions.

THEOREM 2.6. *Let X be a complex projective manifold of dimension $n \leq 4$ with a nef cotangent bundle. Then*

$$\chi(X, K_X) = (-1)^n \cdot \chi(X, \mathcal{O}_X) \geq 0. \tag{2.5}$$

PROOF. The case of curves and surfaces was already considered in Proposition 2.4.

Assume $\dim X = 3$. The nonnegativity of Schur polynomials of Ω_X^1 implies:

$$P_{(3,0,0)}(c) = c_3 \geq 0, \; P_{(2,1,0)}(c) = c_1c_2 - c_3 \geq 0, \; P_{(1,1,1)}(c) = c_1^3 - 2c_1c_2 + c_3 \geq 0.$$

Then, by using Riemann–Roch, we have

$$\chi(X, \Omega_X^3) = -\chi(X, \mathcal{O}_X) = \frac{c_1c_2}{24} \geq 0.$$

Assume next that $\dim X = 4$. The nonnegativity of Schur polynomials of Ω^1_X implies:

$$P_{(4,0,0,0)}(c) = c_4 \geq 0,$$
$$P_{(3,1,0,0)}(c) = c_1c_3 - c_4 \geq 0,$$
$$P_{(2,2,0,0)}(c) = c_2^2 - c_1c_3 \geq 0,$$
$$P_{(2,1,1,0)}(c) = c_1^2c_2 - c_1c_3 - c_2^2 + c_4 \geq 0,$$
$$P_{(1,1,1,1)}(c) = c_1^4 - 3c_1^2c_2 + 2c_1c_3 + c_2^2 - c_4 \geq 0.$$

A calculation via the Riemann–Roch theorem yields the following:

$$\chi(X, \Omega^4_X) = \chi(X, \mathcal{O}_X) = \frac{-c_1^4 + 4c_1^2c_2 + c_1c_3 + 3c_2^2 - c_4}{720}.$$

Since c_1^4 has a negative coefficient in $\chi(X, \Omega^4_X)$, the nonnegativity of the Schur polynomials is not sufficient to guarantee that $\chi(X, \Omega^4_X) \geq 0$. However, since Ω^1_X is nef by our assumptions, it follows that K_X is nef (hence X is minimal), so the corresponding Miyaoka–Yau type inequality [**41, 47**] yields:

$$c_1^4 \leq \frac{5}{2}c_1^2c_2. \tag{2.6}$$

We then have:

$$\begin{aligned}
-c_1^4 + 4c_1^2c_2 + c_1c_3 + 3c_2^2 - c_4 &\geq -c_1^4 + 4c_1^2c_2 + c_1c_3 + 3c_2^2 - c_4 - P_{(1,1,1,1)}(c)\\
&= -2c_1^4 + 7c_1^2c_2 - c_1c_3 + 2c_2^2\\
&\geq -2c_1^4 + 7c_1^2c_2 - c_1c_3 + 2c_2^2 - 2P_{(2,2,0,0)}(c)\\
&= -2c_1^4 + 7c_1^2c_2 + c_1c_3\\
&\geq -2c_1^4 + 7c_1^2c_2 + c_1c_3 - P_{(3,1,0,0)}(c)\\
&= -2c_1^4 + 7c_1^2c_2 + c_4\\
&\geq -2c_1^4 + 7c_1^2c_2 + c_4 - P_{(4,0,0,0)}(c)\\
&= -2c_1^4 + 7c_1^2c_2\\
&\geq \frac{4}{5}c_1^4,
\end{aligned}$$

where the last inequality follows by (2.6). Finally, we have by [**12**, Corollary 2.6] that $c_1^4 \geq 0$. Altogether, we get that $\chi(X, \Omega^4_X) = \chi(X, \mathcal{O}_X) \geq 0$. □

2.2. On Conjecture 1.3. In this section, we prove the following variant of Conjecture 1.3. We refer to [**38, 39**] for background material on mixed Hodge modules.

THEOREM 2.7. *Let X be a complex projective manifold with a finite morphism $f\colon X \to A$ to an abelian variety, and let $\mathcal{M}$ be a mixed Hodge module on X. Then for any integer p one has:*

$$\chi(X, Gr^p_F DR(\mathcal{M})) \geq 0, \tag{2.7}$$

where $Gr^p_F DR(\mathcal{M})$ are the graded pieces, with respect to the Hodge filtration, of the de Rham complex associated to $\mathcal{M}$.

PROOF. Since f is a finite morphism, $\mathcal{N} := f_*\mathcal{M} = f_!\mathcal{M}$ is a mixed Hodge module on the abelian variety A. And since f is proper and the functor $Gr_F^p DR$ commutes with proper pushforward, we have:

$$\chi(X, Gr_F^p DR(\mathcal{M})) = \chi(A, Gr_F^p DR(\mathcal{N})).$$

Finally, by [**37**, Corollary 2.7], there exists $L \in Pic^0(A)$ such that

$$H^i(A, Gr_F^p DR(\mathcal{N}) \otimes L) = 0, \text{ for all } i \neq 0.$$

Therefore,

$$\chi(A, Gr_F^p DR(\mathcal{N})) = \chi(A, Gr_F^p DR(\mathcal{N}) \otimes L) \geq 0,$$

which proves the assertion. □

2.3. On Conjecture 1.6. In this section we discuss various aspects of Conjecture 1.6.

First, it was shown in [**12**, Proposition 3.10] that, if TX is nef, then one of the two situations occurs:

(i) $c_1(X)^n = c_1(TX)^n = 0$, in which case it follows that all Chern polynomials of X of weighted degree $2n$ vanish as well (cf. [**12**, Corollary 2.7]). In particular, by Riemann–Roch, $\chi(X, \Omega^p) = 0$ for all p, and Conjecture 1.6 is true in this case.

(ii) $c_1(X)^n > 0$, in which case X is a Fano manifold (i.e., a complex projective manifold with K_X^{-1} ample). Then the Kodaira vanishing theorem implies that

$$H^q(X, \mathcal{O}_X) = H^q(X, K_X \otimes K_X^{-1}) = 0, \quad \text{for } q \geq 1.$$

Hence, $\chi^0(X) = \chi(X, \mathcal{O}_X) = \dim H^0(X, \mathcal{O}_X) = 1$. By Serre duality, we further get that $(-1)^n \chi^n(X) > 0$.

It therefore remains to prove Conjecture 1.6 in the Fano case for $0 < p < n$. In what follows, we prove this case of Conjecture 1.6 under the additional assumption that X has a cellular decomposition, in the sense that there is a chain of Zariski closed subsets $X_i \subseteq X$ such that $X_i \setminus X_{i+1}$ is a union of affine spaces. More precisely, we have the following.

PROPOSITION 2.8. *If the complex projective manifold X has a cellular decomposition (e.g., X is rational homogenous), then $(-1)^p \cdot \chi^p(X) > 0$ for all $p \leq \dim X$.*

PROOF. Since X has a cellular decomposition, the cycle map $cl_X : A^*(X) \to H^*(X)$ is an isomorphism; see [**15**, Example 19.1.11(b)]. In particular, the cohomology of X is generated by algebraic cycles. Therefore the Hodge numbers of X are concentrated on the diagonal, i.e., $h^{p,q} = 0$ unless $p = q$. Furthermore, it follows by Hard Lefschetz and the fact that $h^{0,0} = 1$ that $h^{p,p} > 0$ for all $p \leq \dim X$. (An alternative proof of this fact follows along the lines of [**3**, Theorem 3.1], using induction over the number of cells together with the fact that $H_c^i(\mathbb{C}^k; \mathbb{Q}) = 0$ for $i \neq 2k$ and $H_c^{2k}(\mathbb{C}^k; \mathbb{Q}) = \mathbb{C}$ is pure of weight $2k$.) Therefore,

$$(-1)^p \cdot \chi(X, \Omega^p) = h^{p,p} > 0.$$

□

Altogether, we obtain the following.

THEOREM 2.9. *If X is a compact Kähler manifold of dimension n with nonnegative bisectional curvature (e.g., nonnegative sectional curvature), then for $0 \leq p \leq n$ one has*

$$(-1)^p \cdot \chi^p(X) \geq 0, \tag{2.8}$$

with $\chi^p(X) := \chi(X, \Omega_X^p)$.

PROOF. First recall that the bisectional curvature can be written as a positive linear combination of two sectional curvatures (e.g., see [**48**]). So if X has nonnegative sectional curvature, then it also has a nonnegative bisectional curvature.

Secondly, our assumptions imply that the tangent bundle TX is nef. Indeed, a compact Kähler manifold X with nonnegative bisectional curvature has (by definition) a tangent bundle TX which is Griffiths semipositive. It is also known that Griffiths semipositive bundles are nef, cf. [**12**].

As above, it suffices to further restrict our attention to the case when X is a Fano manifold. Then X is simply connected, and hence, by Mok's theorem [**31**, Main Theorem], X is a product of complex projective spaces and compact Hermitian symmetric spaces (of rank ≥ 2), all of which are known to admit a cellular decomposition. Hence X admits a cellular decomposition, and the claim follows now by Proposition 2.8. □

REMARK 2.10. By results of Carrell and Liebermann [**9**, Theorem 1], an alternative approach to proving the diagonal concentration of Hodge numbers, and thus to prove Conjectures 1.9 and 1.6, is to show that X admits a holomorphic vector field with only isolated zeros. Note that, if X is as in (*ii*) above, then Bott's residue formula implies that any vector field on X has a nonempty vanishing locus. Moreover, if TX is globally generated, Bertini's theorem tells us that the zero locus of a general vector field on X consists of only isolated points.

REMARK 2.11. As already mentioned in the introduction, Proposition 2.8 can be seen as reducing the proof of Conjecture 1.6 to the Campana–Peternell Conjecture 1.7. This also applies to the situation described in the previous remark, since rational homogeneous spaces are known to carry holomorphic vector fields with only isolated zeros. Moreover, while the nefness of the tangent bundle implies by (2.1) the nonnegativity of the top Segre class $s_n(X) := s_n(TX)$, it is known that rational homogeneous spaces have positive top Segre class [**11**, Theorem 4.1]. This motivates the reduction in [**11**] of Conjecture 1.7 to the nonvanishing of the top Segre class.

Acknowledgment

We thank the anonymous referees for carefully reading the manuscript and for making several constructive suggestions.

References

[1] M. Albanese, L. F. Di Cerbo, and L. Lombardi, *Aspherical complex surfaces, the Singer conjecture, and Gromov-Lück inequality* $\chi \geq |\sigma|$, `arXiv:2311.10226 [v3 []`, (2024).

[2] D. Arapura and B. Wang, *Perverse sheaves on varieties with large fundamental groups*, Journal of Diff. Geom. (to appear).

[3] A. Björner and T. Ekedahl, *On the shape of Bruhat intervals*, Ann. of Math. (2) **170** (2009), no. 2, 799–817, DOI 10.4007/annals.2009.170.799. MR2552108

[4] S. Bloch and D. Gieseker, *The positivity of the Chern classes of an ample vector bundle*, Invent. Math. **12** (1971), 112–117, DOI 10.1007/BF01404655. MR297773

[5] A. Borel and J. Tits, *Compléments à l'article: "Groupes réductifs"* (French), Inst. Hautes Études Sci. Publ. Math. **41** (1972), 253–276. MR315007

[6] F. Campana and T. Peternell, *Projective manifolds whose tangent bundles are numerically effective*, Math. Ann. **289** (1991), no. 1, 169–187, DOI 10.1007/BF01446566. MR1087244

[7] J. Cao and F. Xavier, *Kähler parabolicity and the Euler number of compact manifolds of non-positive sectional curvature*, Math. Ann. **319** (2001), no. 3, 483–491, DOI 10.1007/PL00004444. MR1819879

[8] J. B. Carrell, *Holomorphic* $\mathbf{C}^*$ *actions and vector fields on projective varieties*, Topics in the theory of algebraic groups, Notre Dame Math. Lectures, vol. 10, Univ. Notre Dame Press, Notre Dame, IN, 1982, pp. 1–37. MR698785

[9] J. B. Carrell and D. I. Lieberman, *Holomorphic vector fields and Kaehler manifolds*, Invent. Math. **21** (1973), 303–309, DOI 10.1007/BF01418791. MR326010

[10] S.-s. Chern, *On curvature and characteristic classes of a Riemann manifold*, Abh. Math. Sem. Univ. Hamburg **20** (1955), 117–126, DOI 10.1007/BF02960745. MR75647

[11] J.-P. Demailly, M. Păun, and T. Peternell, *Fano manifolds with big and nef tangent bundles*, Preprint, (2015).

[12] J.-P. Demailly, T. Peternell, and M. Schneider, *Compact complex manifolds with numerically effective tangent bundles*, J. Algebraic Geom. **3** (1994), no. 2, 295–345. MR1257325

[13] P. Eyssidieux, *La caractéristique d'Euler du complexe de Gauss-Manin* (French, with English summary), J. Reine Angew. Math. **490** (1997), 155–212, DOI 10.1515/crll.1997.490.155. MR1468929

[14] W. Fulton and R. Lazarsfeld, *Positive polynomials for ample vector bundles*, Ann. of Math. (2) **118** (1983), no. 1, 35–60, DOI 10.2307/2006953. MR707160

[15] W. Fulton, *Intersection theory*, 2nd ed., Ergebnisse der Mathematik und ihrer Grenzgebiete. 3. Folge. A Series of Modern Surveys in Mathematics [Results in Mathematics and Related Areas. 3rd Series. A Series of Modern Surveys in Mathematics], vol. 2, Springer-Verlag, Berlin, 1998, DOI 10.1007/978-1-4612-1700-8. MR1644323

[16] M. Gromov, *Kähler hyperbolicity and* L_2*-Hodge theory*, J. Differential Geom. **33** (1991), no. 1, 263–292. MR1085144

[17] H.-L. Gu, *A new proof of Mok's generalized Frankel conjecture theorem*, Proc. Amer. Math. Soc. **137** (2009), no. 3, 1063–1068, DOI 10.1090/S0002-9939-08-09611-1. MR2457447

[18] *Guido's book of conjectures*, Enseign. Math. (2) **54** (2008), no. 1-2, 3–189. A gift to Guido Mislin on the occasion of his retirement from ETHZ, June 2006; Collected by Indira Chatterji. MR2454697

[19] J.-M. Hwang, *Holomorphic vector fields with totally degenerate zeroes*, Compositio Math. **105** (1997), no. 1, 65–77, DOI 10.1023/A:1000181820373. MR1436745

[20] F. E. A. Johnson and D. Kotschick, *On the signature and Euler characteristic of certain four-manifolds*, Math. Proc. Cambridge Philos. Soc. **114** (1993), no. 3, 431–437, DOI 10.1017/S0305004100071711. MR1235990

[21] J. Jost and K. Zuo, *Vanishing theorems for* L^2*-cohomology on infinite coverings of compact Kähler manifolds and applications in algebraic geometry*, Comm. Anal. Geom. **8** (2000), no. 1, 1–30, DOI 10.4310/CAG.2000.v8.n1.a1. MR1730897

[22] A. Kanemitsu, *Fano 5-folds with nef tangent bundles*, Math. Res. Lett. **24** (2017), no. 5, 1453–1475, DOI 10.4310/MRL.2017.v24.n5.a6. MR3747171

[23] A. Kanemitsu, *Extremal rays and nefness of tangent bundles*, Michigan Math. J. **68** (2019), no. 2, 301–322, DOI 10.1307/mmj/1549681299. MR3961218

[24] J. Kollár, *Shafarevich maps and automorphic forms*, M. B. Porter Lectures, Princeton University Press, Princeton, NJ, 1995, DOI 10.1515/9781400864195. MR1341589

[25] J. Konarski, *On a special case of Carrell's conjecture*, J. Algebra **182** (1996), no. 1, 38–44, DOI 10.1006/jabr.1996.0159. MR1388855

[26] R. Lazarsfeld, *Positivity in algebraic geometry. II*, Ergebnisse der Mathematik und ihrer Grenzgebiete. 3. Folge. A Series of Modern Surveys in Mathematics [Results in Mathematics and Related Areas. 3rd Series. A Series of Modern Surveys in Mathematics], vol. 49, Springer-Verlag, Berlin, 2004. Positivity for vector bundles, and multiplier ideals, DOI 10.1007/978-3-642-18808-4. MR2095472

[27] Q. Li, *Smooth projective horospherical varieties with nef tangent bundles*, Indiana Univ. Math. J. **66** (2017), no. 6, 1879–1893, DOI 10.1512/iumj.2017.66.6265. MR3744812

[28] Y. Liu, L. Maxim, and B. Wang, *Aspherical manifolds, Mellin transformation and a question of Bobadilla-Kollár*, J. Reine Angew. Math. **781** (2021), 1–18, DOI 10.1515/crelle-2021-0055. MR4343101

[29] L. Maxim, *On singular generalizations of the Singer-Hopf conjecture*, Math. Nachr. **296** (2023), no. 11, 5232–5241. MR4674275

[30] Y. Miyaoka, *On the Chern numbers of surfaces of general type*, Invent. Math. **42** (1977), 225–237, DOI 10.1007/BF01389789. MR460343

[31] N. Mok, *The uniformization theorem for compact Kähler manifolds of nonnegative holomorphic bisectional curvature*, J. Differential Geom. **27** (1988), no. 2, 179–214. MR925119

[32] S. Mori, *Projective manifolds with ample tangent bundles*, Ann. of Math. (2) **110** (1979), no. 3, 593–606, DOI 10.2307/1971241. MR554387

[33] R. Muñoz, G. Occhetta, L. E. Solá Conde, K. Watanabe, and J. A. Wiśniewski, *A survey on the Campana-Peternell conjecture*, Rend. Istit. Mat. Univ. Trieste **47** (2015), 127–185, DOI 10.13137/0049-4704/11223. MR3456582

[34] R. Muñoz, G. Occhetta, L. E. Solá Conde, and K. Watanabe, *Rational curves, Dynkin diagrams and Fano manifolds with nef tangent bundle*, Math. Ann. **361** (2015), no. 3-4, 583–609, DOI 10.1007/s00208-014-1083-x. MR3319542

[35] R. Nomura, *Miyaoka-Yau inequality for compact Kähler manifolds with semi-positive canonical bundles*, J. Geom. Anal. **31** (2021), no. 10, 9633–9638, DOI 10.1007/s12220-021-00622-3. MR4303936

[36] G. Occhetta, L. E. Solá Conde, K. Watanabe, and J. A. Wiśniewski, *Fano manifolds whose elementary contractions are smooth $\mathbb{P}^1$-fibrations: a geometric characterization of flag varieties*, Ann. Sc. Norm. Super. Pisa Cl. Sci. (5) **17** (2017), no. 2, 573–607. MR3700378

[37] M. Popa and C. Schnell, *Generic vanishing theory via mixed Hodge modules*, Forum Math. Sigma **1** (2013), Paper No. e1, 60, DOI 10.1017/fms.2013.1. MR3090229

[38] M. Saito, *Modules de Hodge polarisables* (French), Publ. Res. Inst. Math. Sci. **24** (1988), no. 6, 849–995 (1989), DOI 10.2977/prims/1195173930. MR1000123

[39] M. Saito, *Mixed Hodge modules*, Publ. Res. Inst. Math. Sci. **26** (1990), no. 2, 221–333, DOI 10.2977/prims/1195171082. MR1047415

[40] Y. T. Siu and S. T. Yau, *Compact Kähler manifolds of positive bisectional curvature*, Invent. Math. **59** (1980), no. 2, 189–204, DOI 10.1007/BF01390043. MR577360

[41] H. Tsuji, *Stability of tangent bundles of minimal algebraic varieties*, Topology **27** (1988), no. 4, 429–442, DOI 10.1016/0040-9383(88)90022-5. MR976585

[42] K. Watanabe, *Fano 5-folds with nef tangent bundles and Picard numbers greater than one*, Math. Z. **276** (2014), no. 1-2, 39–49, DOI 10.1007/s00209-013-1185-2. MR3150191

[43] K. Watanabe, *Fano manifolds of coindex three admitting nef tangent bundle*, Geom. Dedicata **210** (2021), 165–178, DOI 10.1007/s10711-020-00538-2. MR4200914

[44] S. T. Yau, *Calabi's conjecture and some new results in algebraic geometry*, Proc. Nat. Acad. Sci. U.S.A. **74** (1977), no. 5, 1798–1799, DOI 10.1073/pnas.74.5.1798. MR451180

[45] S. T. Yau, *Problem section*, Seminar on Differential Geometry, Ann. of Math. Stud., No. 102, Princeton Univ. Press, Princeton, NJ, 1982, pp. 669–706. MR645762

[46] Q. Zhang, *Global holomorphic one-forms on projective manifolds with ample canonical bundles*, J. Algebraic Geom. **6** (1997), no. 4, 777–787. MR1487236

[47] Y. Zhang, *Miyaoka-Yau inequality for minimal projective manifolds of general type*, Proc. Amer. Math. Soc. **137** (2009), no. 8, 2749–2754, DOI 10.1090/S0002-9939-09-09838-4. MR2497488

[48] F. Zheng, *Complex differential geometry*, AMS/IP Studies in Advanced Mathematics, vol. 18, American Mathematical Society, Providence, RI; International Press, Boston, MA, 2000, DOI 10.1090/amsip/018. MR1777835

Department of Mathematics, Purdue University, West Lafayette, Indiana 47907
Email address: `arapura@purdue.edu`

Department of Mathematics, University of Wisconsin-Madison, 480 Lincoln Drive, Madison, Wisconsin 53706-1388
Email address: `maxim@math.wisc.edu`

Department of Mathematics, University of Wisconsin-Madison, 480 Lincoln Drive, Madison, Wisconsin 53706-1388
Email address: `wang@math.wisc.edu`

Contemporary Mathematics
Volume **816**, 2025
https://doi.org/10.1090/conm/816/16354

On Lipschitz cohomology of aspherical manifolds

Alexander Dranishnikov

Abstract. We introduce the notion of Lipschitz cohomology classes of a group with local coefficients and reduce the Novikov conjecture for a group Γ to the question whether the Berstein–Schwarz class $\beta_\Gamma \in H^1(\Gamma, I(\Gamma))$ is Lipschitz.

1. Introduction

The Novikov higher signature conjecture states that the higher signatures of closed oriented manifolds are invariant under orientation-preserving homotopy equivalences. For aspherical manifolds it is equivalent to say that the rational Pontryagin classes are homotopy invariant. The conjecture has a long and eventful history (see the survey [**Yu**]) and it is still open.

One of the approaches to the Novikov conjecture was introduced more than 30 years ago by Connes, Gromov, and Moscovici by defining Lipschitz cohomology classes of a discrete group and proving the Novikov conjecture for such classes [**CGM**]. Their approach recovered the Novikov conjecture for all classes of groups where it was proven by that time. In particular, it gave another proof of the Novikov conjecture for hyperbolic groups. More than 20 years ago T. Kato used Lipschitz cohomology to prove the Novikov conjecture for the combable groups [**K1**],[**K2**]. In [**Dr**] Lipschitz cohomology were used to recover Carlsson–Pedersen result [**CP**] on the Novikov conjecture for groups Γ admitting certain equivariant compactification of the classifying space $E\Gamma$.

Since then the Lipschitz cohomology approach to the Novikov conjecture seems to be completely abandoned. This paper is an attempt to revive it.

Lipschitz cohomology classes of a discrete group were defined in [**CGM**] as the collection of images under a family of certain slant product homomorphisms denoted $\alpha_\cap$ and defined by maps $\alpha : \Gamma \times P \to \mathbb{R}^n$ where P is a parameter space. Connes, Gromov, and Moscovici first defined a light version of such classes called *proper Lipschitz cohomology*. It turns out that in all cases where all cohomology classes of an aspherical manifold are known to be proper Lipshitz, like in the case of manifolds with nonpositive curvature $K(M) \leq 0$, they all belong to the image of one homomorphism $\alpha_\cap$. In such situations we say that the group Γ has *canonically Lipschitz cohomology*. Unfortunately, the canonically Lipschitz cohomology

2020 *Mathematics Subject Classification.* Primary 20J06; Secondary 57N65.

Key words and phrases. Lipschitz cohomology, aspherical manifolds, Berstein–Schwarz cohomology class.

The author was supported in part by Simons Foundation.

cannot lead to a proof of the Novikov conjecture, since it was shown in [**Dr**] that there are aspherical manifolds which do not have all its rational cohomology classes canonically Lipschitz.

It was noticed in [**CGM**] that to prove that the 0-dimensional cohomology class $1 \in H^0(\Gamma, \mathbb{R}) = \mathbb{R}$ is proper Lipschitz could be a difficult problem. In the second half of the paper [**CGM**] the authors introduced a heavier version of Lipschitz cohomology classes for which the Novikov conjecture still holds true and for which is quite easy to show that the class $1 \in H^0(\Gamma, \mathbb{R})$ is Lipschitz for any group.

In this paper we explore the possibility of proving the Novikov conjecture for a group Γ by proving that its Berstein–Schwarz class $\beta_\Gamma \in H^1(\Gamma, I(\Gamma))$ is Lipschitz. Here $I(\Gamma)$ is the augmentation ideal. Due to universality of β_Γ [**DR**],[**Sc**] every cohomology class of Γ is the image of some power of β_Γ under a coefficient homomorphism. In order to use this property of β_Γ with the aim the Novikov conjecture we extend the notion of Lipschitz cohomology classes to cohomology with local coefficients. Like in [**CGM**] we do it in two steps: First we define proper Lipschitz cohomology and then general Lipschitz cohomology.

We note that, since the authors in [**CGM**] worked with differential forms, their definition of Lipschitz cohomology classes makes sense only for real coefficients. In this paper we use a combinatorial approach. For that, first we give an integral version of Connes–Gromov–Moscovici definition of proper Lipschitz cohomology with constant coefficients and then we transform it to a definition of proper Lipschitz cohomology with coefficients in a $\mathbb{Z}\Gamma$-module. We prove the product theorem, the coefficient homomorphism theorem, and the connecting homomorphism theorem for proper Lipschitz cohomology with local coefficients. These results allow us to reduce the Novikov conjecture to the question about proper Lipschitz property of the 0-dimensional class $1 \in H^0(\Gamma, \mathbb{Z}) = \mathbb{Z}$. As it was already noted above, this is far from being easy.

In [**CGM**] the definition of proper Lipschitz cohomology classes of a group Γ was extended by dropping the properness of the Γ-action on the parameter space P. We extend our definition to the nonproper case similarly. The bad news about general Lipschitz cohomology with local coefficients is that the connecting homomorpphism theorem does not hold anymore. It means that we cannot reduce the Novikov conjecture to the above question about 0-dimensional cohomology. The good news is that the product theorem and the coefficient homomorphism theorem still hold true. In view of universality of β_Γ we can reduce the Novikov conjecture to the question about 1-dimensional cohomology. Namely, we reduce it to the question whether the Berstein–Schwarz class β_Γ is Lipschitz.

Also in the paper we computed the cohomology group $H^k(\Gamma, I(\Gamma)^{\otimes k}) = \mathbb{Z}$ in the case when Γ is the fundamental group of a closed aspherical orientable manifold. Thus, since the Novikov conjecture is rational, to prove it for a group Γ it suffices to find any nonzero Lipschitz cohomology class in $H^1(\Gamma, I(\Gamma))$.

The paper is organized as follows. In Section 2 we give a combinatorial definition of proper Lipschitz cohomology with constant coefficients and present examples of aspherical manifolds with canonically proper Lipschitz integral cohomology. In Section 3 we define the Berstein–Schwarz cohomology class, state the universality theorem, and carry out computations of cohomology of aspherical manifolds with coefficients in $I(\Gamma)^{\otimes k}$. In Section 4 first we give definition of proper and then general Lipschitz cohomology with local coefficients. There we prove the coefficient homomorphism, the product, and the connecting homomorphism theorems.

1.1. Some notations. We denote cohomology of a group Γ with coefficients in $\mathbb{Z}\Gamma$-module L as $H^n(\Gamma, L)$ and cohomology of a space X with $\pi_1(X) = \Gamma$ as $H^n(X; L)$. Thus, $H^n(B\Gamma; L) = H^n(\Gamma, L)$. For $\mathbb{Z}\Gamma$-module L and M the tensor product $L \otimes M$ means the tensor product over $\mathbb{Z}$ and $L \otimes_\Gamma M$ means the tensor product over $\mathbb{Z}\Gamma$. The tensor product of n copies of L over $\mathbb{Z}$ will be denoted as $L^{\otimes n}$. The group of covariants of L is denoted by L_Γ.

2. Proper Lipschitz cohomology of groups

In [**CGM**] Connes, Gromov, and Moscovici defined proper Lipschitz cohomology with coefficients in $\mathbb{R}$. In this section we present a combinatorial definition which works for any ring, in particular for $\mathbb{Z}$ and $\mathbb{Q}$.

2.1. Equivariant slant product. We recall the slant product. Let C and D be chain complexes [**Sp**]:

$$/ : Hom(C \otimes D, \mathbb{Z}) \times D \to Hom(C, \mathbb{Z})$$

is defined by the equality $(\phi/c')(c) = \phi(c \otimes c')$. Thus, if $\phi \in Hom((C \otimes D)_n, \mathbb{Z})$ and $c' \in D_k$, then $\phi/c' \in Hom(C_{n-k}, \mathbb{Z})$. This slant product defines the slant product for cohomology

$$/ : H^n(C \otimes D) \times H_k(D) \to H^{n-k}(C).$$

Now we assume that the group Γ acts on the chain complexes C. Then there is a Γ-equivariant version of the slant product

$$/ : Hom_\Gamma(C \otimes D, \mathbb{Z}) \times D \to Hom_\Gamma(C, \mathbb{Z})$$

where $Hom_\Gamma(C, \mathbb{Z})$ denotes the group of Γ-invariant homomorphisms.

Suppose that the discrete group Γ acts freely on a finite dimensional locally finite simplicial complex P by simplicial transformations. Let C be a free resolution of $\mathbb{Z}$ over $\mathbb{Z}\Gamma$ and let $D = C(P)^\Gamma$ to be the chain complex that consist of Γ-invariant infinite simplicial chains on P. Thus, D is isomorphic to the chain complex of infinite simplicial chains $C^{inf}(P/\Gamma)$ on the orbit sapce.

We consider the equivariant slant product

$$H^n_\Gamma(C \otimes D) \times H_k(P : \Gamma) \to H^{n-k}_\Gamma(C) = H^{n-k}(\Gamma)$$

where $H_k(P : \Gamma) = H_k(C^{inf}(P/\Gamma))$. Note that if P/Γ is compact, then $H_k(P : \Gamma) = H_k(\Gamma)$.

2.2. Proper Lipschitz cohomology classes. The proper Lipschitz cohomology classes of a discrete group Γ were defined by Connes, Gromov, and Moscovici [**CGM**] by means of the following data: A finite-dimensional connected locally finite simplicial complex P with a free simplicial Γ-action and a map $\alpha : \Gamma \times P \to \mathbb{R}^n$ satisfying the conditions

(1) α is invariant with respect to the diagonal action of Γ on $\Gamma \times P$;
(2) the restriction $\alpha \mid_{\gamma \times P}$ is proper for all $\gamma \in \Gamma$;
(3) the restriction $\alpha \mid_{\Gamma \times p}$ is 1-Lipschitz for all $p \in P$ for the word metric on Γ.

Since all word metrics on Γ generated by finite sets are quasi-isometric, rescaling metric on $\mathbb{R}^n$ shows that condition (3) is independent of choice of the word metric on Γ.

Let D be as above and let $C = C_*(\Delta^\infty(\Gamma))$ be the simplicial chain complex of the infinite simplex spanned by Γ. Given the above map $\alpha : \Gamma \times P \to \mathbb{R}^n$ we define an extension $\bar{\alpha} : \Delta^\infty(\Gamma) \times P \to \mathbb{R}^n$ by means of the barycentric coordinates

$$\bar{\alpha}((\sum_{i=0}^{k} t_i\gamma_i) \times x) = \sum_{i=0}^{k} t_i \alpha(\gamma_i \times x).$$

Let ω be a singular cocycle generating $H^n_c(\mathbb{R}^n)$ with the support in some ball $B \subset \mathbb{R}^n$. It turns out to be the cochain $\bar{\alpha}^*(\omega)$ is well defined on infinite chains $C(\Gamma) \otimes D$. Let $\sum_\sigma n_\sigma$ be an infinite ℓ-dimensional simplicial integral chain on P where the sum is taken over all ℓ-simplices in P. Let Δ be a simplex in $\Delta^\infty(\Gamma)$ of dimension $k = n - \ell$. Then by the definition

$$\bar{\alpha}^*(\omega)(\Delta \otimes \sum_\sigma n_\sigma \sigma) = \sum_\sigma n_\sigma \omega(\bar{\alpha}|(\Delta \times \sigma))$$

where the restriction of $\bar{\alpha}$ to $\Delta \times \sigma$ is denoted as $\bar{\alpha}|(\Delta \times \gamma\sigma)$ and is treated as the singular n-chain on $\mathbb{R}^n$. Here we use the staircase triangulation [**L**] of the product of oriented simplices $\Delta \times \sigma$ to define the simplicial chain.

By virtue of condition (2) the map $\bar{\alpha} : \Delta \times P \to \mathbb{R}^n$ is proper on $y \times P$ for each vertex $y \in \Delta$ and by condition (3) the image of $\Delta \times x$ is uniformly bounded on $x \in P$. Therefore, the restriction of $\bar{\alpha}$ to $\Delta \times P$ is proper. Hence there is a compact subset $D \subset P$ such that $\bar{\alpha}(\Delta \times (P \setminus D)) \cap B = \emptyset$. Since D intersects only finitely many simplices σ, the sum above is finite. Thus, $\bar{\alpha}^*(\omega)$ is well defined.

If the chain $z = \sum_\sigma n_\sigma$ is Γ-invariant, the value $\bar{\alpha}^*(\omega)(\Delta \otimes \hat{\sigma})$ is Γ-invariant on the first factor:

$$\bar{\alpha}^*(\omega)(\gamma'\Delta \otimes z) = \sum_{\hat{\sigma} \subset P/\Gamma} \sum_{\gamma \in \Gamma} n_\sigma \omega(\bar{\alpha}|(\gamma'\Delta \times \gamma\sigma))$$

$$= \sum_{\hat{\sigma} \subset P/\Gamma} \sum_{\gamma \in \Gamma} n_\sigma \omega(\bar{\alpha}|(\Delta \times (\gamma')^{-1}\gamma\sigma) = \bar{\alpha}^*(\omega)(\Delta \otimes z).$$

Here we split the infinite sum $z = \sum_\sigma n_\sigma = \sum_{\hat{\sigma}} \sum_{\gamma \in \Gamma} n_\sigma(\gamma\sigma)$ into the sums along different orbits $\hat{\sigma} = \Gamma\sigma$ of simplices.

Note that $\bar{\alpha}^*(\omega)$ is cocycle being the image of a cocycle. Let $[\bar{\alpha}^*(\omega)]$ denote the cohomology class of the cocycle $\bar{\alpha}^*(\omega)$.

We define the map $\alpha_\cap : H_k(P : \Gamma) \to H^{n-k}(\Gamma)$ as

$$\alpha_\cap(a) = [\bar{\alpha}^*(\omega)]/a.$$

DEFINITION 2.1. All classes in the image $\alpha_\cap(H_*(P : \Gamma)) \subset H^*(\Gamma)$ are called integral *proper Lipschitz cohomology classes* of a group Γ.

PROBLEM 2.2. *Is the 0-dimensional cohomology class* $1 \in H^0(\Gamma, \mathbb{Z})$ *proper Lipschitz for every group* Γ*?*

In the rest of the paper we will try to convince the reader that this is an important question in the Lipschitz cohomology theory (see Corollary 4.9).

The above definition can be done in terms of a cellular Γ-action on P and it can be done directly with use of $\bar{\alpha}$. What we need for that is the finiteness condition of the sum $\sum_{\gamma \in \Gamma} \omega(\bar{\alpha}|(\Delta \times \gamma\sigma))$. Suppose that a CW complex $B\Gamma$ is given a metric that agrees with its topology on every finite subcomplex. We consider the induced metric and CW structure on the universal cover $E\Gamma$.

PROPOSITION 2.3. *Let $B \subset \mathbb{R}^n$ be the unit ball centered at 0. Suppose that a map $\bar{\alpha} : E\Gamma \times P \to \mathbb{R}^n$ satisfies the conditions*

(1) *$\bar{\alpha}$ is invariant with respect to the diagonal action of Γ on $\Gamma \times P$;*
(2) *the restriction $\bar{\alpha} \mid_{x\times P}$ is proper for all $x \in E\Gamma$;*
(3) *the restriction $\bar{\alpha} \mid_{E\Gamma\times y}$ is 1-Lipschitz for all $y \in P$.*

Then for every cells e in $E\Gamma$ and c in P the intersection $\bar{\alpha}(e \times \gamma c) \cap B = \emptyset$ for all but finitely many $\gamma \in \Gamma$.

PROOF. We fix $y_0 \in P$. Suppose that $diam(e) \leq d$, for $d > 1$. For $x \in e$ in view of the propernes of map $\bar{\alpha}|_{x\times P} : x \times P \to \mathbb{R}^n$ there is a compact neighborhood D of y_0 such that $\bar{\alpha}(x \times (P \setminus D)) \subset \mathbb{R}^n \setminus B(0, 2d)$. Since by condition (3), $\bar{\alpha}(e \times y)$ has diameter $\leq d$ and we obtain that $\bar{\alpha}(e \times (P \setminus D)) \subset \mathbb{R}^n \setminus B(0, d)$. The properness of the action of Γ on P implies that only finitely many translates of c lie in D. □

EXAMPLE 2.4. The space of probability measures on Γ with finite supports $P_f(\Gamma)$ can be identified with $\Delta^\infty(\Gamma)$. We consider the Wasserstein metric $W_1(\mu, \nu)$ on $P_f(\Gamma)$ which due to the Kantorovich-Rubinstein duality can be defined by the formula

$$W_1(\mu, \nu) = \sup_{f \in Lip^1(\Gamma,\mathbb{R})} |\int f d\mu - \int f d\nu|$$

where $Lip^1(X, \mathbb{R})$ denote the set of 1-Lipschitz real-valued functions on X. We note that the inclusion $\Gamma \subset P_f(\Gamma)$ is an isometric embedding. Then any 1-Lipschitz map $g : \Gamma \to Z$ defines a 1-Lipschitz map $P(g) : P_f(\Gamma) \to P_f(Z)$. Since the barycenter map $b_{\mathbb{R}} : P_f(\mathbb{R}) \to \mathbb{R}$ defined as $b(\mu) = \int t d\mu$ is 1-Lipschitz, the composition $b_{\mathbb{R}^n} \circ P(\alpha) : P_f(\Gamma) \to \mathbb{R}^n$ with the barycenter map $b_{\mathbb{R}^n} : P_f(\mathbb{R}^n) \to \mathbb{R}^n$ is $\sqrt{n}$-Lipschitz.

The map $\bar{\alpha} : \Delta^\infty(\Gamma) \times P \to \mathbb{R}^n$ defined as

$$\bar{\alpha}(x, y) = \frac{1}{\sqrt{n}} b_{\mathbb{R}^n} \circ P(\alpha|_{\Gamma\times y})(x, y)$$

satisfies conditions (1)–(3).

PROPOSITION 2.5. *Suppose that a map $\alpha : \Gamma \times P \to \mathbb{R}^n$ satisfies conditions (1)–(3) for a group Γ with a finite CW complex $B\Gamma$ and a finite-dimensional simplicial complex P. Then there is a map $\bar{\alpha} : E\Gamma \times P \to \mathbb{R}^n$, where $E\Gamma$ is the universal cover of $B\Gamma$, satisfying the conditions*

(1) *α is invariant with respect to the diagonal action of Γ on $E\Gamma \times P$;*
(2) *the restriction $\alpha \mid_{y\times P}$ is proper for all $y \in E\Gamma$;*
(3) *the restriction $\alpha \mid_{E\Gamma\times p}$ is 1-Lipschitz for all $p \in P$ for the metric on $E\Gamma$ lifted from $B\Gamma$.*

PROOF. For the proof we use the Michael selection theorem.

THEOREM 2.6 ([**M**]). *Let X be a paracompact space and Y a Banach space. Let $F : X \to Y$ be a lower semi-continuous set-valued map with nonempty closed convex values. Then there exists a continuous selection of F.*

The map $\alpha = (\alpha_j)$ is defined by its coordinate maps $\alpha_j : \Gamma \times P \to \mathbb{R}$. We construct an extension $\bar{\alpha}_j : E\Gamma \times P \to \mathbb{R}$ for each j. We may assume that the CW complex $B\Gamma$ has one vertex. Then the 0-skeleton of $E\Gamma$ can be identified with Γ. We construct the map $\bar{\alpha}$ recursively defining it on $E\Gamma^{(i)} \times P$ for the i-skeleton of $E\Gamma$. The base of induction is the map α_j.

Suppose that $\bar{\alpha}_j : E\Gamma^{(i)} \times P \to \mathbb{R}$ satisfying (1) and (3) is constructed. We extend it separately to $\bar{e}^{i+1} \times P$ for each $(i+1)$-dimensional cell $e^{i+1} \subset E\Gamma$. Let $Lip^1(X, \mathbb{R})$ denote the set of 1-Lipschitz functions. Let (X, A) be a compact pair and let $f_0 : A \to \mathbb{R}$ be a 1-Lipschitz function.

Claim: Let (X, A) be a compact pair and let $f_0 : A \to \mathbb{R}$ be a 1-Lipschitz map.

(i) *Then the set* $Lip^1_{A,f_0}(X, \mathbb{R})$ *of 1-Lipschitz functions* $g : X \to \mathbb{R}$ *that restrict to* f_0 *on* A, $g|_A = f_0$, *is a nonempty compact convex set in the Banach space* $C(X, \mathbb{R})$.

(ii) *If a map* $f : P \to Lip^1(A, \mathbb{R})$ *is continuous, then the multi-valued map* $F : P \to C(X, \mathbb{R})$ *defined as* $F(p) = Lip^1_{A,f(p)}(X, \mathbb{R})$ *is a lower semi-continuous.*

Proof of the Claim: Since a 1-Lipschitz map to $\mathbb{R}$ can be extended to a 1-Lipschitz map for any pair (X, A) [**CGM**], the set is nonempty. Note that the straight-line homotopy in $\mathbb{R}$ between 1-Lipschitz maps as a path that consists of 1-Lipschitz maps. This proves convexity (i). Compactness follows from the Ascoli–Arzela theorem.

Let $p_k \to p$ be a sequence in P converging to p. Let $g_k \in F(p_k)$, $g_k : X \to \mathbb{R}$. Going to a subsequence in view of the Ascoli–Arzela theorem we may assume that g_k converges to a function $g : X \to \mathbb{R}$. Note that g is 1-Lipschitz and $g|_A = f(p)$. Hence, $g \in F(p)$. This implies the semi-continuity of F.

We apply Claim (ii) with $X = \bar{e}^{i+1}$, $A = \partial e^{i+1}$, and $f : P \to Lip^1(A, \mathbb{R})$ obtained from $\bar{\alpha}_j : A \times P \to \mathbb{R}$ to obtain a lower semi-continuous multi-valued function $F : P \to C(X, \mathbb{R})$ defined as $F(p) = Lip^1_{A,f(p)}(X, \mathbb{R})$. By the Michael selection theorem there is a continuous selection $\phi : P \to C(X, \mathbb{R})$. The map ϕ defines the desired extension $\bar{\alpha}$ to $\bar{e}^{i+1} \times P$.

We do this for all $(i+1)$-dimensional cells to get $\bar{\alpha}_j$ defined on $E\Gamma^{(i+1)} \times P$. Since we assume that the metric on $E\Gamma^{(i+1)}$ is geodesic, the local 1-Lipschitz condition extends to global. Thus condition (3) is satisfied.

As the result we obtain an extension $\bar{\alpha}_j : E\Gamma \times P \to \mathbb{R}$ of α_j that satisfies (1) and (3). Define $\bar{\alpha} = (\bar{\alpha}_j)_{j=1}^n : E\Gamma \times P \to \mathbb{R}^n$. We rescale $\mathbb{R}^n$ to have condition (3).

We need to check condition (2), i.e., to show that $\bar{\alpha}$ is proper on $y \times P$ for each $y \in E\Gamma$. Suppose that $|\bar{\alpha}(y \times p_n)| \leq R$. Then

$$|(\alpha(\gamma \times p_n)| \leq |\bar{\alpha}(y \times p_n)| + |(\alpha(\gamma \times p_n) - \bar{\alpha}(y \times p_n)| \leq R + d(y, \gamma)$$

for fixed γ. Since $\alpha|_{\gamma \times P}$ is proper, this implies that the sequence p_n is bounded in P. □

Since the map $\bar{\alpha}$ is invariant under the diagonal action of Γ, there is a map $\hat{\alpha} : E\Gamma \times_\Gamma P \to \mathbb{R}^n$ such that $\bar{\alpha} = \hat{\alpha}\bar{p}$ where $\bar{p} : E\Gamma \times P \to E\Gamma \times_\Gamma P$ is the projection to the orbit space. Then $\bar{\alpha}^*(\omega) = \bar{p}^*\hat{\alpha}^*(\omega)$. In the case of finite $B\Gamma$ the map $\hat{\alpha}$ is proper. Therefore,

$$[\bar{\alpha}^*(\omega)] = \bar{p}^*(\hat{\alpha}^*([\omega]))$$

where $\hat{\alpha}^*([\omega]) \in H^n_c(E\Gamma \times_\Gamma P)$. Thus, the cohomology class $[\bar{\alpha}^*(\omega)]$ is a Γ-invariant version of $\hat{\alpha}^*([\omega])$.

2.3. Manifolds with $K \leq 0$. Let M be a closed parallelizable Rimannian n-manifold with nonpositive sectional curvature and let $\Gamma = \pi_1(M)$. Then M is $B\Gamma$. Let $q : TM \to \mathbb{R}^n$ be a trivialization map of the tangent bundle which is an isomorphism on each fiber. By the Cartan–Hadamard theorem the exponential map $exp_x : T_xM \to M$ lifts to a diffeomorphism $\widetilde{exp}_x = exp_y : T_xM \to \widetilde{M}$, where $p : \widetilde{M} \to M$ is the universal covering amd $y \in p^{-1}(x)$. The collection of maps $\widetilde{exp}_x$, $x \in M$ defines a fiberwise diffeomorphism

$$\widetilde{exp} : TM \to \widetilde{M} \times_\Gamma \widetilde{M}.$$

We note that the inverse map $\widetilde{exp_x}^{-1}$ is 1-Lipschitz. Let $\Lambda : \widetilde{M} \times_\Gamma \widetilde{M} \to TM$ be the inverse for $\widetilde{exp}$. Here the projection of $TM \to M$ corresponds to the projection $\widetilde{M} \times_\Gamma \widetilde{M} \to M$ onto the second factor. We treat the first factor as $E\Gamma$. Consider the map

$$\bar{\alpha} = q \circ \Lambda \circ \bar{p} : E\Gamma \times \widetilde{M} \to \mathbb{R}^n.$$

The map $\hat{\alpha} = q\Lambda$ is proper, since q is proper and Λ is a homeomorphism. Since the restriction $\bar{p}|_{x\times\widetilde{M}} : x \times \widetilde{M} \to E\Gamma_\Gamma\widetilde{M}$, it follows that the restriction $\bar{\alpha}|_{x\times\widetilde{M}}$ is proper for all $x \in E\Gamma$. Since $\widetilde{exp_x}^{-1}$ is 1-Lipschitz, the restriction $\bar{\alpha}|_{E\Gamma\times y}$ is 1-Lipschitz for all y. Thus, conditions (1)–(3) are satisfied.

We note that $q^*([\omega])$ is the Thom class u of the tangent bundle $TM \to M$ which is supported on a tubular neighborhood of the zero section of TM. Then $\hat{\alpha}^*([\omega]) = \Lambda^*(u)$ is the Thom class of the bundle $E\Gamma \times_\Gamma \widetilde{M} \to M$ which is supported on a tubular neighborhood V of the "zero" section, i.e., the image $\bar{p}(\Delta\widetilde{M})$ of the diagonal. We may assume that the neighborhood V is pulled back from the tubular neighborhood of the diagonal $\Delta M \subset M \times M$ under the natural projection $p' : \widetilde{M} \times_\Gamma \widetilde{M} \to M \times M$. Thus, $\hat{\alpha}^*([\omega])$ is the image under $(p')^*$ of the Thom class u' of normal bundle of the diagonal in $M \times M$.

We recall that the Thom class u' of the normal bundle of the diagonal $\Delta M \subset M \times M$ generates the diagonal class $u'' \in H^n(M \times M)$ and the slant product with u'' is the Poincaré duality isomorphism $H_k(M) \to H^{n-k}(M)$. The proof of this fact can be derived from the fomula for u'' given in [**MS**]. This isomorphism can be stated in terms of Γ-equivariant homology and cohomology of $\widetilde{M}$ as

$$(\bar{p}p')^*(u'')/ : H_k^\Gamma(\widetilde{M}) \to H^\Gamma_{n-k}(\widetilde{M}).$$

Thus, the slant product $(\bar{p}p')^*(u')/ : H_k^\Gamma(\widetilde{M}) \to H^\Gamma_{n-k}(\widetilde{M})$ is an isomorphism. Since $(p')^*(u') = \hat{\alpha}^*([\omega])$, the slant product $\alpha_\cap$ is an isomorphism.

Therefore, all cohomology of Γ are proper Lipschitz [**CGM**].

If M is not parallelizable the same result can be achieved by adding to TM the complementary vector bundle and repeating the above argument with a large parameter space P. Then one should use the Poincaré duality for the infinite chain homology (sometimes called the Borel–Moore homology).

2.4. The case of aspherical manifolds. Now we consider the case when $B\Gamma = M$ is a closed orientable aspherical n-manifold and $E\Gamma = \widetilde{M}$ is its universal cover, $\Gamma = \pi_1(M)$. Suppose that a map $\alpha : \Gamma \times P \to \mathbb{R}^n$ satisfies conditions (1)–(3). Then by Proposition 2.5 there is a map $\bar{\alpha} : \widetilde{M} \times P \to \mathbb{R}^n$ satisfying conditions (1)–(3). First, we consider the case when $P = \widetilde{M}$. If the map $\hat{\alpha} : \widetilde{M} \times_\Gamma P \to \mathbb{R}^n$ induces an injective homomorphism of the n-dimensional cohomology with compact supports, then $\hat{\alpha}^*([\omega])$ is the Thom class of the normal bundle of $M = P/\Gamma \subset \widetilde{M} \times_\Gamma P$ and the argument in the example in Section 2.3 implies that $\alpha_\cap$ is an isomorphism. It means that all cohomology classes of Γ are proper Lipschitz.

One can expand the case when $\alpha_\cap$ is an isomorphism for aspherical manifolds as follows. Suppose that Γ acts on $P = \widetilde{M} \times \mathbb{R}^k$ in such a way that the projection $pr : P \to \widetilde{M}$ is equivariant. Then for rational coefficients the argument with the Thom class allows to prove the following proposition.

PROPOSITION 2.7 ([**Dr**]). *Let $\alpha : \widetilde{M} \times P \to \mathbb{R}^{n+k}$ be a map satisfying conditions* (1)–(3). *Then $\alpha_\cap$ is isomorphism over $\mathbb{Q}$ in all dimensions if and only if $\alpha_\cap$ is not zero.*

If the conditions of Proposition 2.7 are satisfied for an aspherical manifold M we call the cohomology groups $H^*(M;\mathbb{Q}) = H^*(\pi_1(M),\mathbb{Q})$ *canonically proper Lipschitz.* In particular, Proposition 2.7 allows to extend the argument of Proposition 2.3 to not necessarily parallelizable manifolds with sectional curvature $K \leq 0$. We refer to [**CGM**] for detailed account. Briefly, if E is a $\mathbb{R}^k$-bundle "orthogonal" to the topological tangent bundle $\widetilde{M} \times_\Gamma \widetilde{M} \cong TM$ of M, then the Whitney sum admits a trivialization $q : TM \oplus E \to \mathbb{R}^{n+k}$. The pull-back $E^* \to \widetilde{M}$ is a trivial vector bundle and it is Γ-equivariant, hence, $P = \widetilde{M} \times \mathbb{R}^k$ satisfies the condition of Proposition 2.7. Note that the orbit space of the diagonal action on $\widetilde{M} \times P$ is the total space of the bundle $(\widetilde{M} \times_\Gamma \widetilde{M}) \oplus E$ and hence we can use the Thom class to show essentiality of $\hat{\alpha}$.

One of the corollaries of Proposition 2.7 states that if an aspherical manifold M has a proper Lipschitz class in some dimension (say, dimension 0) with the above P then all its cohomology classes are proper Lipschitz.

The following theorem about general aspherical manifolds [**Dr**] is a generalization of the case of manifolds with $K(M) \leq 0$.

THEOREM 2.8. *Suppose that the universal cover of an aspherical manifold M with $\pi_1(M) = \Gamma$ admits a Γ-equivariant Čech acyclic Higson dominated compactification, then M has canonically proper Lipschitz cohomology and, hence, the Novikov conjecture holds true for Γ.*

We recall that the Higson domination condition for a compactification of $E\Gamma$ is equivalent to the nullity condition from the definition of Bestvina's Z-boundary [**Best**].

In particular, this theorem implies the Novikov conjecture for the fundamental group of an aspherical manifold $\Gamma = \pi_1(M)$ when Γ admits an EZ-boundary. We refer to [**FL**] for the original proof of the Novickov conjecture for groups admitting EZ-boundary.

3. Berstein–Schwarz class of aspherical manifolds

3.1. Berstein–Schwarz class. Let Γ be a discrete group, $\mathbb{Z}\Gamma$ denote the group ring, and let $\epsilon : \mathbb{Z}\Gamma \to \mathbb{Z}$ be the augmentation homomorphism. Then the augmentation ideal $I(\Gamma)$ is the kernel of ϵ. The Berstein–Schwarz class $\beta_\Gamma \in H^1(\Gamma, I(\Gamma))$ is the image $\beta_\Gamma = \delta(1)$ of the generator $1 \in H^0(\Gamma,\mathbb{Z})$ under connecting homomorphism $\delta : H^0(\Gamma,\mathbb{Z}) \to H^1(\Gamma, I(\Gamma))$ in the coefficient long exact sequence generated by the short exact sequence

$$0 \to I(\Gamma) \to \mathbb{Z}\Gamma \stackrel{\epsilon}{\to} \mathbb{Z} \to 0.$$

We recall [**Br**] that the cap product $\alpha_1\alpha_2$ of two cohomology classes $\alpha_i \in H^{k_i}(\Gamma, L_i)$, $i = 1, 2$, with coefficients in $\mathbb{Z}\Gamma$-modules L_i belongs to $H^{k_1+k_2}(\Gamma, L_1 \otimes L_2)$ where the tensor product is taken over $\mathbb{Z}$.

THEOREM 3.1 (Universality). [**DR**],[**Sc**] *For any $\mathbb{Z}\Gamma$-module L and any cohomology class $\alpha \in H^k(\Gamma, L)$ there is a homomorphism $\phi : I(\Gamma)^{\otimes k} \to L$ such that α is the image of $(\beta_\Gamma)^k$ under the induced homomorphism*

$$\phi^* : H^k(\Gamma, I(\Gamma)^{\otimes k}) \to H^k(\Gamma, L).$$

In the case when Γ is the fundamental group of a closed orientable aspherical manifold the group $H^k(\Gamma, I(\Gamma)^{\otimes k})$ can be computed.

THEOREM 3.2. *For a closed orientable aspherical n-manifold M with the fundamental group Γ for $k < n$ the cohomology group $H^k(M; I(\Gamma)^{\otimes k})$ is an infinite cyclic group generated by $(\beta_\Gamma)^k$.*

The cohomology group $H^n(M; I(\Gamma)^{\otimes n})$ is the group of coinvariants $(I(\Gamma)^{\otimes n})_\Gamma = I(\Gamma)^{\otimes n} \otimes_\Gamma \mathbb{Z}$.

PROOF. For any ℓ the short exact sequences

$$0 \to I(\Gamma) \to \mathbb{Z}\Gamma \to \mathbb{Z} \to 0$$

defines by tensor product with $I(\Gamma)^{\otimes \ell-1}$ a short exact sequence

$$0 \to I(\Gamma)^{\otimes \ell} \to I(\Gamma)^{\otimes \ell-1} \otimes \mathbb{Z}\Gamma \to I(\Gamma)^{\otimes \ell-1} \to 0.$$

We note that the Γ-module $I(\Gamma)^{\otimes \ell-1} \otimes \mathbb{Z}\Gamma$ is projective [**DR**]. Then the coefficient long exact sequence for homology and the fact that homology of a group with coefficients in projective module are trivial in dimensions > 0 [**Br**] produce the isomorphisms for $i > 0$

$$H_{i+1}(M; I(\Gamma)^{\otimes \ell-1}) \to H_i(M; I(\Gamma)^{\otimes \ell}).$$

In particular,

$$\mathbb{Z} = H_n(M; \mathbb{Z}) = H_{n-1}(M; I(\Gamma)) = H_{n-2}(M; I(\Gamma)^{\otimes 2}) = \cdots = H_1(M; I(\Gamma)^{\otimes n-1}).$$

By the Poincare duality with local coefficients [**Br**]

$$H^k(M; I(\Gamma)^{\otimes k}) = H_{n-k}(M; I(\Gamma)^{\otimes k}) = \mathbb{Z}$$

for $k < n$ and

$$H^n(M; I(\Gamma)^{\otimes n}) = H_0(\Gamma, I(\Gamma)^{\otimes n}) = (I(\Gamma)^{\otimes n})_\Gamma. \qquad \square$$

PROPOSITION 3.3. *For any group π there is an isomorphism $H_1(\pi, \mathbb{Z}) = I(\pi)_\pi$. For $\ell > 1$ there is a short exact sequence*

$$0 \to H_\ell(\pi; \mathbb{Z}) \to (I(\pi)^{\otimes \ell})_\pi \to I(\pi)I(\pi)^{\otimes \ell-1} \to 0.$$

PROOF. The equality $H_1(\pi, \mathbb{Z}) = I(\pi)_\pi$ follows from the coefficients long exact sequence

$$0 = H_1(\pi, \mathbb{Z}\pi) \to H_1(\pi, \mathbb{Z}) \to H_0(\pi, I(\pi)) \to H_0(\pi, \mathbb{Z}\pi) \to H_0(\pi, \mathbb{Z}) \to 0$$

and the fact that $H_0(\pi, \mathbb{Z}\pi) = (\mathbb{Z}\pi)_\pi \to \mathbb{Z} = H_0(\pi, \mathbb{Z})$ is an isomorphism. For $\ell > 1$ we have the following exact sequence

$$0 \to H_1(\pi, I(\pi)^{\otimes \ell-1}) \to H_0(\pi, I(\pi)^{\otimes \ell}) \to H_0(\pi, I(\pi)^{\otimes \ell-1} \otimes \mathbb{Z}\pi)$$
$$\to H_0(\pi, I(\pi)^{\otimes \ell-1}) \to 0,$$

which defines the short exact sequence

$$0 \to H_1(\pi, I(\pi)^{\otimes \ell-1}) \to H_0(\pi, I(\pi)^{\otimes \ell}) \to K_{\ell-1} \to 0$$

where

$$K_\ell = \mathrm{Ker}\{(I(\pi)^{\otimes \ell} \otimes \mathbb{Z}\pi)_\pi \xrightarrow{1\otimes\epsilon} (I(\pi)^{\otimes \ell} \otimes \mathbb{Z})_\pi\}.$$

In view of the equality $(M \otimes N)_\pi = M \otimes_\pi N$ we obtain

$$K_\ell = Ker\{I(\pi)^{\otimes \ell} \to (I(\pi)^{\otimes \ell})_\pi\} = I(\pi)I(\pi)^{\otimes \ell}.$$

The chain of isomorphisms

$$H_{i+1}(\pi, I(\pi)^{\otimes k-1}) \to H_i(\pi; I(\pi)^{\otimes k})$$

for $i \geq 1$ defines the equalities

$$H_1(\pi, I(\pi)^{\otimes \ell-1}) = H_2(\pi, I(\pi)^{\otimes \ell-2}) = \cdots = H_{l-1}(\pi, I(\pi)) = H_\ell(\pi, \mathbb{Z}). \qquad \square$$

THEOREM 3.4. *For a closed orientable aspherical manifold M with the fundamental group Γ,*

$$H^n(M; I(\Gamma)^{\otimes n}) = I(\Gamma)I(\Gamma)^{\otimes n-1} \oplus \mathbb{Z}.$$

PROOF. By Proposition 3.3 there is the short exact sequence

$$0 \to \mathbb{Z} \xrightarrow{i} H^n(M; I(\Gamma)^{\otimes n}) \xrightarrow{j} I(\Gamma)I(\Gamma)^{\otimes n-1} \to 0.$$

As it follows from the commutative diagram

$$\begin{array}{ccccccc} H_n(M;\mathbb{Z}) & \xrightarrow{\cap\beta_\Gamma} & H_{n-1}(M.I(\Gamma)) & \xrightarrow{\cap\beta_\Gamma} & \dots & \xrightarrow{\cap\beta_\Gamma} & H_0(M; I(\Gamma)^{\otimes n}) \\ {\scriptstyle [M]\cap}\uparrow & & {\scriptstyle [M]\cap}\uparrow & & & & {\scriptstyle [M]\cap}\uparrow \\ H^0(M;\mathbb{Z}) & \xrightarrow{\cup\beta_\Gamma} & H^1(M; I(\Gamma)) & \xrightarrow{\cup\beta_\Gamma} & \dots & \xrightarrow{\cup\beta_\Gamma} & H^n(M; I(\Gamma)^{\otimes n}) \end{array}$$

the homomorphism i being the Poincare dual to the iterated cup product with β_Γ is an iterated $\cap$-product with β_Γ. Hence $i : H_n(M;\mathbb{Z}) \to H_0(M; I(\Gamma)^{\otimes n})$ takes the fundamental class $[M]$ to $[M] \cap (\beta_\Gamma)^n$, the Poincaré dual to $(\beta_\Gamma)^n$. By the universality of the Berstein–Schwarz class there is a homomorphism of Γ-modules $\phi : I(\Gamma)^{\otimes n} \to \mathbb{Z}$ such that the induced homomorphism

$$\phi^* : H^n(M; I(\Gamma)^{\otimes n}) \to H^n(M, \mathbb{Z}) = \mathbb{Z}$$

takes $(\beta_\Gamma)^n$ to the generator $1 \in \mathbb{Z}$. Let $H^n(M;\mathbb{Z}) = \mathbb{Z} \xrightarrow{\psi} \mathbb{Z} = H_n(M;\mathbb{Z})$ be the isomorphism that takes 1 to $[M]$. We obtain the commutative diagram

$$\begin{array}{ccc} H^n(M;\mathbb{Z}) & \xleftarrow{\phi^*} & H^n(M; I(\Gamma)^{\otimes n}) \\ {\scriptstyle \psi}\downarrow & & {\scriptstyle [M]\cap}\downarrow \\ H_n(M;\mathbb{Z}) & \xrightarrow{-\cap(\beta_\Gamma)^n} & H_0(M; I(\Gamma)^{\otimes n}) \end{array}$$

that defines the splitting. $\square$

4. Lipschitz cohomology with local coefficients

4.1. Proper Lipschitz cohomology with local coefficients. Let Γ be a discrete group with finite classifying complex $B\Gamma$. We recall that the proper Lipschitz cohomology classes of Γ were by the following data [**CGM**]. Suppose that there is a finite-dimensional connected complex P with a proper Γ-action and a map $\alpha : \Gamma \times P \to \mathbb{R}^n$ satisfying the conditions

(1) α is invariant with respect to the diagonal action of Γ on $\Gamma \times P$;
(2) the restriction $\alpha\mid_{\gamma \times P}$ is proper for all $\gamma \in \Gamma$;
(3) the restriction $\alpha\mid_{\Gamma \times p}$ is 1-Lipschitz for all $p \in P$ for the word metric on Γ.

Connes, Gromov, and Moscovici used constant real coefficients $\mathbb{R}$. Here we present the definition which extends to local coefficients L with base ring $\mathbb{Z}$ as well as $\mathbb{R}$.

We assume that the action of Γ on P is a free simplicial action. Let $D = C_*(P)$ be the simplicial chain complex of P and let S_m denote the set of all m-simplices in P. The group of Γ-invariant m-chains $D^\Gamma_m(L) = (D_m \otimes L)^\Gamma$ with coefficients in L consists of formal sums $\sum_{\sigma \in S_m} \lambda_\sigma \sigma$ where coefficients $\lambda_\sigma \in L$ are satisfying the equalities $\lambda_{\gamma\sigma} = \gamma\lambda_\sigma$.

PROPOSITION 4.1. *If D is a finitely generated $\mathbb{Z}\Gamma$-module, then $D^\Gamma(L) \cong L \otimes_\Gamma D$.*

PROOF. Since D is finitely generated, its every element $\sum_{\sigma \in S_m} \lambda_\sigma \sigma$ can be split as a finite sum $\sum_i \sum_{\gamma \in \Gamma} \lambda_{\gamma\sigma_i}(\gamma\sigma_i)$. The map $\Phi : D^\Gamma(L) \to L \otimes_\Gamma D$ defined on generators as

$$\Phi(\sum_\gamma \lambda_{\gamma\sigma}(\gamma\sigma)) = \lambda_\sigma \otimes_\Gamma \sigma$$

is a well-defined homomorphism, since

$$\lambda_\sigma \otimes_\Gamma \sigma = \gamma\lambda_\sigma \otimes_\Gamma \gamma\sigma = \lambda_{\gamma\sigma} \otimes_\Gamma \gamma\sigma.$$

It is an isomorphism with the inverse Φ^{-1} that takes $\lambda \otimes_\Gamma \sigma$ to $\sum_\gamma \gamma\lambda(\gamma\sigma)$. □

Let C_* be the normal resolution of Γ, i.e., the simplicial chain complex of the simplex $\Delta^\infty(\Gamma)$ spanned by Γ [**Br**]. Suppose that $\phi : (C \otimes D)_n \to \mathbb{Z}$ is a Γ-invariant cocycle such that the sum

$$\sum_\sigma \phi(c \otimes \sigma)$$

is finite for any Γ-invariant infinite chain $z = \sum_\sigma n_\sigma \sigma$ in $D^\Gamma = D^\Gamma(\mathbb{Z})$ where σ runs over all $(n-k)$-dimensional simplices in P. Thus, this sum defines a Γ-invariant cocycle $\bar{\phi} : (C \otimes D^\Gamma)_n \to \mathbb{Z}$. We define a Γ-invariant slant product

$$\bar{\phi}/ : D^\Gamma_{n-k}(L) \to Hom_\Gamma(C_k, L)$$

by the formula

$$(\bar{\phi}/z)(\Delta) = \sum_{\sigma \in S_{n-k}} \phi(\Delta \otimes \sigma)\lambda_\sigma$$

where $z = \sum_{\sigma \in S_{n-k}} \lambda_\sigma \sigma$ is a Γ-invariant $(n-k)$-chain with coefficients in L and $\Delta = [\gamma_0, \dots, \gamma_k]$ is a k-simplex in $\Delta^\infty(\Gamma)$.

Next, we show that the cochain $\bar{\phi}/z : C_k \to L$ is equivariant:

$$\begin{aligned}(\bar{\phi}/z)(g\Delta) &= \sum_{\sigma} \phi(g\Delta \otimes \sigma)\lambda_\sigma \\ &= \sum_{\sigma} \phi(\Delta \otimes g^{-1}\sigma) g g^{-1}\lambda_\sigma \\ &= \sum_{\sigma} \phi(\Delta \otimes g^{-1}\sigma) g \lambda_{g^{-1}\sigma} \\ &= g(\bar{\phi}/z)(\Delta).\end{aligned}$$

Here we use the fact that $g^{-1} : S_{n-k} \to S_{n-k}$ is a bijection. The cochain $\bar{\phi}/z$ is a cocycle when $z = \sum_{\gamma \in \Gamma} \lambda_{\gamma\sigma} \gamma\sigma$ is a Γ-invariant cycle. Indeed, it is a cocycle since $\bar{\phi}$ is a cocycle and applying the Leibnitz formula $\partial(c \otimes a) = \partial c \otimes a \pm c \otimes \partial a$ we obtain $\delta(\bar{\phi}/z) = (\delta\bar{\phi})/z \pm \phi/\partial z = 0$.

Then we show that slant product with $\bar{\phi}$ defines the homomorphism

$$H_{n-k}(D^{\Gamma}_*(L)) \to H^k(C_*; L)$$

which we denote as $[\bar{\phi}]/$. Let z be a Γ-invariant cycle as above and let $z' = \sum_{\sigma} \mu_\sigma \sigma'$ be an invariant $(n-k+1)$-dimensional chain. Then $\bar{\phi}/(z + \partial z') = \bar{\phi}/z + \bar{\phi}/\partial z'$. Note that $\bar{\phi}/\partial z' = \delta(\bar{\phi}/z')$. Therefore, $\bar{\phi}/(z + \partial z')$ defines the same cohomology class as $\bar{\phi}/z$.

Assume that we have a map $\alpha : \Gamma \times P \to \mathbb{R}^n$ satisfying conditions (1)–(3) and let $\bar{\alpha} : \Delta^{\infty}(\Gamma) \times \Gamma \to \mathbb{R}$ be the extension defined in Section 2. Let ω be a cocycle on $\mathbb{R}^n$ whose cohomology class $[\omega]$ generates the group $H^n_c(\mathbb{R}^n; \mathbb{Z}) = \mathbb{Z}$. We define $\phi = \bar{\alpha}^*(\omega)$. The homomorphism $\phi : C \otimes D|_n \to \mathbb{Z}$ is defined by the formula

$$\phi(\Delta \otimes \sigma) = \omega(\bar{\alpha}|\Delta \times \sigma)$$

where we treat $\bar{\alpha} : \Delta^k \times \sigma^{n-k} \to \mathbb{R}^n$ as a singular chain with the staircase triangulation of the product of two oriented simplices. The cocycle ϕ is Γ-invariant, since so is the map α. The argument in Section 2.2 shows that the sum

$$\sum_{\sigma} \phi(c \otimes \sigma)$$

is finite.

We denote by $H_k(P : \Gamma; L)$ the homology group defined by infinite invariant chains with coefficients in a Γ-module L. Note that if P/Γ is compact $H_k(P : \Gamma; L) = H_k(P/\Gamma; L)$. As in the case of constant coefficients [**CGM**] we denote the corresponding slant product $[\bar{\phi}]/$ homomorphism by

$$\alpha_{\cap} : H_k(P : \Gamma; L) \to H^{n-k}(\Gamma, L).$$

Thus, the cohomology class $\alpha_{\cap}([\sum_{\sigma} \lambda_\sigma \sigma])$ for a Γ-invariant cycle $\sum_{\sigma} \lambda_\sigma \sigma$ is defined by the cocycle

$$[\gamma_0, \dots, \gamma_k] \mapsto \sum_{\sigma} \omega(\bar{\alpha}|[\gamma_0, \dots, \gamma_k] \times \sigma)\lambda_\sigma.$$

DEFINITION 4.2. All classes in the image $\alpha_{\cap}(H_*(P/\Gamma; L)) \subset H^*(\Gamma, L)$ are called *proper Lipschitz cohomology classes* of a group Γ with coefficients in the $\mathbb{Z}\Gamma$-module L.

REMARK 4.3. In the case of constant coefficients $L = \mathbb{R}$ the above formula recovers the Connes–Gromov–Moscovici definition of Lipschitz cohomology as it presented in Section 2.

REMARK 4.4. In the above construction we can use cellular chain complexes C_* and D_* whenever $E\Gamma$ and P are given Γ-equivariant CW-complex structures.

THEOREM 4.5. *The image of a proper Lipschitz class under coefficients homomorphism is proper Lipschitz.*

PROOF. Let $h : L \to L'$ be a homomorphism of $\mathbb{Z}\Gamma$-modules and let $a = [\bar{\phi}]/[z] \in H^k(\Gamma, L)$ be a proper Lipschitz cohomology class realized by a map $\alpha : \Gamma \times P \to \mathbb{R}^n$. Then $h^*(a) = [\bar{\phi}]/h_*([b])$. Indeed, if $z = \sum_\sigma \lambda_\sigma \sigma$, then $h_*(z) = \sum_\sigma h(\lambda_\sigma)\sigma$ and

$$(\bar{\phi}/h_*(z))(\Delta^k) = \sum_\sigma \phi(\Delta^k \otimes \sigma)h(\lambda_\sigma) = h^*(\bar{\phi}/b)(\Delta^k)). \qquad \square$$

Given a short exact sequence of $\mathbb{Z}\Gamma$-modules $0 \to K \to L \xrightarrow{\phi} M \to 0$ and a chain complex of free $\mathbb{Z}\Gamma$-modules D_* the sequence $0 \to D_*^\Gamma(K) \to D_*^\Gamma(L) \to D_*^\Gamma(M) \to 0$ is exact. For finitely generated $\mathbb{Z}\Gamma$-modules D_* it follows from Proposition 4.1. For infinitely generated D_* we leave the proof to the reader.

THEOREM 4.6. *Let $0 \to K \to L \xrightarrow{\phi} M \to 0$ be a short exact sequence of $\mathbb{Z}\Gamma$-modules, C_* and D_* be chain complexes of free $\mathbb{Z}\Gamma$-modules, and let $a^* \in H^n(C \otimes D^\Gamma; \mathbb{Z})$. Then the diagram*

$$\begin{array}{ccc} H_{n-k}(D^\Gamma(M)) & \xrightarrow{a^*/} & H^k(C, M) \\ \bar{\partial}\big\downarrow & & \bar{\delta}\big\downarrow \\ H_{n-k-1}(D^\Gamma(K)) & \xrightarrow{a^*/} & H^{k+1}(C, K) \end{array}$$

commutes up to the sign, where $\bar{\partial}$ and $\bar{\delta}$ are connecting homomorphisms in the coefficient homology and cohomology long exact sequences.

PROOF. Let a denote a cocycle representing a^* and let $z = \sum m_\sigma \sigma$ be a cycle in $D_*^\Gamma(M)$, $m_\sigma \in M$ and $m_{\gamma\sigma} = \gamma m_\sigma$. Then by the snake lemma for homology $\bar{\partial}z = \sum \ell_\sigma \partial\sigma$ with $\phi(\ell_\sigma) = m_\sigma$, $\ell_\sigma \in L$ and $\ell_{\gamma\sigma} = \gamma\ell_\sigma$. Then

$$(a/\bar{\partial}z)(c) = \sum_\sigma a(c \otimes \partial\sigma)\ell_\sigma = \pm \sum_\sigma a(\partial c \otimes \sigma)\ell_\sigma$$

where the second equality is due to the Leibnitz formula and the fact that a is a cocycle, $a(\partial(c \otimes \sigma)) = 0$. For a cochain $\psi : C_k \to M$ we denote by $\bar{\psi}$ a lift $\bar{\psi} : C_k \to L$ with respect to $\phi : L \to M$. Since $(a/z)(c') = \sum_\sigma a(c' \otimes \sigma)m_\sigma$, we can define the lift $\overline{a/z}$ by the formula $(\overline{a/z})(c') = \sum_\sigma a(c' \otimes \sigma)\ell_\sigma$ with ℓ_σ chosen above.

By the snake lemma for cohomology we obtain

$$\bar{\delta}(a/z)(c) = (\overline{a/z})(\partial c) = \sum_\sigma a(\partial c \otimes \sigma)\ell_\sigma = (a/\bar{\partial}z)(c). \qquad \square$$

COROLLARY 4.7. *Suppose that the integral 0-dimensional cohomology class $1 \in H^0(\Gamma)$ is proper Lipschitz. Then the Berstein–Schwarz class $\beta_\Gamma \in H^1(\Gamma, I(\Gamma)$ is proper Lipschitz.*

PROOF. We consider the short exact sequence

$$0 \to I(\Gamma) \to \mathbb{Z}\Gamma \xrightarrow{\epsilon} \mathbb{Z} \to 0$$

defined by the augmentation. Let $\alpha : \Gamma \times P \to \mathbb{R}^n$ be a map satisfying conditions (1)–(3) required for proper Lipschitz property of $1 \in H^0(\Gamma)$ and let $\bar{\phi}$ be the cocycle from the definition of $\alpha_\cap$. Since the action of Γ on P is proper, we have $n > 0$. Thus, $H^0(\Gamma) \ni 1 = \alpha_\cap([z])$ for some $[z] \in H_n(P : \Gamma)$. By Theorem 4.6 we obtain $\alpha_\cap(\bar{\partial}[z]) = \bar{\delta}(1) = \beta_\Gamma$. □

PROPOSITION 4.8. *Suppose that a class $b \in H^k(\Gamma, L)$ is proper Lipschitz via a map $\alpha : \Gamma \times P \to \mathbb{R}^n$. Then b is proper Lipschitz via the map $\alpha \times 1 : \Gamma \times P \times \mathbb{R} \to \mathbb{R}^n \times \mathbb{R} = \mathbb{R}^{n+1}$.*

PROOF. Here we consider the pull-back action of Γ on $P \times R$ and we take $\bar{\alpha} = \alpha \times 1$. Let $s = \sum_{m \in \mathbb{Z}} [m, m+1]$ be the infinite 1-chain for the natural simplicial complex structure on $\mathbb{R}$. Note that s is a cycle. Let ω_0 be a generator of $H^1_c(\mathbb{R})$ supported in 1/4-neighborhood of 1/2. Suppose that $[b] = \alpha_\cap([z])$ for some cycle $z = \sum_\sigma \lambda_\sigma \sigma$. Then

$$\begin{aligned}(b/z)(\Delta) &= \sum_\sigma (\omega(\bar{\alpha}|\Delta \times \sigma)\lambda_\sigma \\ &= \sum_\sigma (\omega \otimes \omega_0)((\bar{\alpha} \times 1)|\Delta \times \sigma \times [0,1])\lambda_\sigma \\ &= \sum_{m \in \mathbb{Z}} \sum_\sigma (\omega \otimes \omega_0)((\bar{\alpha} \times 1)|\Delta \times \sigma \times [m, m+1])\lambda_\sigma \\ &= (b/(z \otimes s))(\Delta).\end{aligned}$$

□

COROLLARY 4.9. *Suppose that the integral 0-dimensional cohomology class $1 \in H^0(\Gamma)$ is proper Lipschitz. Then cohomology classes in $H^k(\Gamma)$ are Lipschitz for all k.*

PROOF. Given i, in view of Proposition 4.8 we may assume that $1 \in H^0(\Gamma)$ is realized via a map $\alpha : \Gamma \times P \to \mathbb{R}^n$ with $n > i$. Let $1 = \alpha_\cap([z_1])$ for $[z_1] \in H_n(P : \Gamma)$. By induction on i we show that $(\beta_\Gamma)^{\otimes i} = \alpha_\cap([z_i])$ where $z_i = \bar{\partial} z_{i-1}$ for $i < n$. The base of induction, $i = 1$ is Corollary 4.7.

Assume that $(\beta_\Gamma)^{\otimes i-1} = \alpha_\cap([z_{i-1}])$. We apply Theorem 4.6 to the short exact sequence

$$0 \to I(\Gamma)^{\otimes i} \to I(\Gamma)^{\otimes i-1} \otimes \mathbb{Z}\Gamma \xrightarrow{1 \otimes \epsilon} I(\Gamma)^{\otimes i-1} \to 0$$

to obtain $\alpha_\cap(\bar{\partial} z_{i-1}) = \pm\bar{\delta}(\beta_\Gamma^{\otimes i-1}) = \beta_\Gamma^{\otimes i}$. Then in view of the universality of β_Γ, by virtue of Proposition 4.5, all k-dimensional cohomology classes of Γ are proper Lipschitz. □

THEOREM 4.10. *The cup product of proper Lipschitz cohomology classes is proper Lipschitz.*

PROOF. Let $\bar{\alpha}_1 : \Delta^\infty(\Gamma) \times P_1 \to \mathbb{R}^{n_1}$ and $\bar{\alpha}_2 : \Delta^\infty(\Gamma) \times P_1 \to \mathbb{R}^{n_2}$ be the maps that realize Lipschitz classes $(\alpha_1)_\cap([z_1]) \in H^{k_1}(\Gamma, L)$ and $(\alpha_2)_\cap([z']) \in H^{k_2}(\Gamma, M)$ with $z = \sum \lambda_\sigma \sigma$ and $z' = \sum \mu_{\sigma'} \sigma'$. We take the product $\Delta^\infty(\Gamma) \times \Delta^\infty(\Gamma)$ with diagonal action of Γ for the classifying space $E\Gamma$ and consider a CW-complex $P = P_1 \times P_2$ with the diagonal Γ-action. We define

$$\bar{\alpha} : (\Delta^\infty(\Gamma) \times \Delta^\infty(\Gamma)) \times (P_1 \times P_2) \to \mathbb{R}^{n_1} \times \mathbb{R}^{n_2} = \mathbb{R}^{n_1+n_2}$$

as $\bar{\alpha}(x_1 \times x_2, y_1 \times y_2) = \bar{\alpha}_1(x_1, y_1) \times \bar{\alpha}_2(x_2, y_2)$. We show that

$$\alpha_\cap([z \otimes z']) = (\alpha_1)_\cap([z])(\alpha_2)_\cap([z']).$$

Let $k = k_1 + k_2$ and $n = n_1 + n_2$. Let $\Delta^i \times \Delta^{k-i}$ be a k-cell in $\Delta^\infty(\Gamma) \times \Delta^\infty(\Gamma)$. The cocycle ϕ from the definition of $\alpha_\cap$ is given by the formula

$$\begin{aligned}\phi((\Delta_1 \times \Delta_2) \otimes (\sigma \times \sigma')) &= (\omega_1 \otimes \omega_2)(\bar{\alpha}|(\Delta_1 \times \Delta_2) \times (\sigma \times \sigma')) \\ &= (\omega_1 \otimes \omega_2)(\bar{\alpha}_1|(\Delta_1 \times \sigma) \times \bar{\alpha}_2|(\Delta_2 \times \sigma')) \\ &= \omega_1(\bar{\alpha}_1|(\Delta_1 \times \sigma_1) \otimes \omega_2(\bar{\alpha}_2|(\Delta_2 \times \sigma_2) \\ &= \phi_1(\Delta_1 \otimes \sigma)\phi_2(\Delta_2 \otimes \sigma').\end{aligned}$$

We compare the cocycles $\bar{\phi}/(z_1 \otimes z_2)$ and $(\bar{\phi}_1/z_1) \cup (\bar{\phi}_2/z_2)$ which are both elements of

$$Hom_\Gamma(C(\Delta^\infty(\Gamma)) \otimes C(\Delta^\infty(\Gamma)), L \otimes M).$$

Since $z_1 \otimes z_2 = \sum_{\sigma,\sigma'}(\lambda_\sigma \otimes \mu_{\sigma'})(\sigma \times \sigma')$, we obtain

$$(\bar{\phi}/z \otimes z')(\Delta_1 \times \Delta_2) = \sum_{\sigma,\sigma'} \phi((\Delta_1 \times \Delta_2) \otimes (\sigma \times \sigma'))\lambda_\sigma \otimes \mu_{\sigma'}.$$

Then we evaluate the cup product cocycle using the fact that when we do it with respect to the product resolution $C_*(\Delta^\infty(\Gamma)) \otimes C_*(\Delta^\infty(\Gamma))$ of Γ the cup product cocycle coincides with the cross product cocycle treated as a Γ-equivariant homomorphism instead of $(\Gamma \times \Gamma)$-equivariant (see [**Br**]). Thus,

$$\begin{aligned}((\bar{\phi}_1/z_1) \cup (\bar{\phi}_2/z_2))(\Delta_1 \times \Delta_2) &= \bar{\phi}_1/(\sum_\sigma \lambda_\sigma \sigma)(\Delta_1) \otimes \bar{\phi}_2/(\sum_{\sigma'} \mu_{\sigma'}\sigma')(\Delta_2) \\ &= \sum_\sigma \phi_1(\Delta_1 \times \sigma)\lambda_\sigma \otimes \sum_{\sigma'} \phi_2(\Delta_2 \times \sigma')\mu_{\sigma'} \\ &= \sum_{\sigma,\sigma'} \phi((\Delta_1 \times \Delta_2) \otimes (\sigma \times \sigma'))\lambda_\sigma \otimes \mu_{\sigma'}.\end{aligned}$$

Therefore, the cocycles $\bar{\phi}/(z_1 \otimes z_2)$ and $(\bar{\phi}_1/z_1) \cup (\bar{\phi}_2/z_2)$ coincide. □

4.2. General Lipschitz cohomology. Now in the definition of Lipschitz cohomology we drop the assumption of properness of the action of Γ on P. We assume that P is a simplicial complex and Γ acts by simplicial homeomorphisms. Still the equivariant infinite chains D^Γ for $D = C_*(P)$ are defined and the homomorphism $\alpha_\cap$ can be defined by the same formula. Thus,

$$(\bar{\phi}/z)(\Delta) = \sum_{\sigma \in S_{n-k}} \phi(\Delta \otimes \sigma)\lambda_\sigma$$

for $\phi(\Delta \otimes \sigma) = \omega(\bar{\alpha}|\Delta \times \sigma)$ and $\alpha_\cap = [\bar{\phi}]$.

The difference is in formulas for $\bar{\phi}/z$ in the case when the chain $z = \sum_{\gamma\in\Gamma} \lambda_{\gamma\sigma}(\gamma\sigma)$ is defined by a single orbit. In the proper cohomology case it is

$$\bar{\phi}/z(\Delta) = \sum_{\gamma\in\Gamma} \omega(\bar{\alpha}|\Delta \times \gamma\sigma)\lambda_{\gamma\sigma}$$

whereas in the general case

$$\bar{\phi}/z(\Delta) = \sum_{\gamma\Gamma_\sigma \in \Gamma/\Gamma_\sigma} \omega(\bar{\alpha}|\Delta \times \gamma\sigma)\lambda_{\gamma\sigma}$$

where Γ_σ is the isotropy group of σ and the sum is indexed by the set Γ/Γ_σ of cosets of Γ_σ.

The following two examples are modifications of examples from [**CGM**].

EXAMPLE 4.11. The integral 0-dimensional cohomology class of Γ is Lipschitz for any group Γ.

PROOF. Consider the case when P is a point with a trivial action of Γ and with the constant map $\alpha : \Gamma \times P \to \mathbb{R}^0$. □

EXAMPLE 4.12. All integral 1-dimensional cohomology classes of any finitely generated group Γ are Lipschitz.

PROOF. Let $f : \Gamma \to \mathbb{Z}$ be a 1-cocycle. We take $P = \mathbb{R}$ with the action of Γ defined as $\gamma(x) = x + f(x)$. Define $\alpha : \Gamma \times \mathbb{R} \to \mathbb{R}$ as $\alpha(\gamma, x) = -\gamma^{-1}(x)$. The axioms (1)–(2) are obviously satisfied. Let S be a finite generating for Γ and let $\lambda = \max\{|f(s) \mid s \in S$. Then

$$|\alpha(\gamma, x) - \alpha(\gamma s, x) = |f(\gamma^{-1}) - f((\gamma s)^{-1})| = |f(\gamma s) - f(\gamma)| = |f(s)| \leq \lambda d_S(\gamma s, \gamma).$$

This implies that α is λ-Lipschitz. We rescale $\mathbb{R}$ accordingly to get α 1-Lipschitz.

Suppose that the generator $\omega \in H^1_c(\mathbb{R})$ has support concentrated in a small neighborhood around $1/2 \in \mathbb{R}$. Le $z = \sum_{m\in\mathbb{Z}} m$ be an infinite simplicial 0-chain with respect to the triangulation of $\mathbb{R}$ where $\mathbb{Z}$ is the 0-skeleton. Then $\alpha_\cap([z])$ is defined by the cocycle $\bar{\phi}/z$ which takes value

$$\begin{aligned}(\bar{\phi}/z)([e,\gamma]) &= \sum_m \omega(\bar{\alpha}|[1,\gamma] \times m)\\ &= \sum_m \omega([\alpha(e,m), \alpha(\gamma, m)])\\ &= \sum_m \omega([-m, f(\gamma) - m]).\end{aligned}$$

Since for $r \in \mathbb{Z}$ only r different integer translates of the interval $[0, r]$ can hit $1/2$, we obtain that $(\bar{\phi}/z)([e,\gamma]) = f(\gamma)$. Therefore, $\alpha_\cap([z]) = [f]$. □

PROBLEM 4.13. *Is every integral 2-dimensional cohomology class of a finitely presented group Lipschitz?*

An affirmative answer to this question would imply the Novikov conjecture for 2-dimesional classes which is already known [**Ma**],[**HS**].

REMARK 4.14. We note that in the case when the action of Γ on a simplicial complex P is not free the sequence $0 \to D^\Gamma_*(K) \to D^\Gamma_*(L) \to D^\Gamma_*(M) \to 0$ is not exact. As the result, Theorem 4.6 does not hold anymore.

The proof of the following theorem coincides with the proof of Proposition 4.10.

THEOREM 4.15. *The product of two Lipschitz cohomology classes is Lipschitz.*

PROBLEM 4.16. *For which groups Γ the Berstein–Schwarz class is Lipschitz?*

The main result of [**CGM**] and the universality theorem together with Theorem 4.5 and Theorem 4.15 imply the Novikov conjecture for such groups.

References

[Best] Mladen Bestvina, *Local homology properties of boundaries of groups*, Michigan Math. J. **43** (1996), no. 1, 123–139, DOI 10.1307/mmj/1029005393. MR1381603

[Be] Israel Berstein, *On the Lusternik-Schnirelmann category of Grassmannians*, Math. Proc. Cambridge Philos. Soc. **79** (1976), no. 1, 129–134, DOI 10.1017/S0305004100052142. MR400212

[Br] Kenneth S. Brown, *Cohomology of groups*, Graduate Texts in Mathematics, vol. 87, Springer-Verlag, New York-Berlin, 1982. MR672956

[CP] Gunnar Carlsson and Erik Kjær Pedersen, *Čech homology and the Novikov conjectures for K- and L-theory*, Math. Scand. **82** (1998), no. 1, 5–47, DOI 10.7146/math.scand.a-13823. MR1634649

[CGM] A. Connes, M. Gromov, and H. Moscovici, *Group cohomology with Lipschitz control and higher signatures*, Geom. Funct. Anal. **3** (1993), no. 1, 1–78, DOI 10.1007/BF01895513. MR1204787

[Dr] A. N. Dranishnikov, *Lipschitz cohomology, Novikov's conjecture, and expanders* (Russian, with Russian summary), Tr. Mat. Inst. Steklova **247** (2004), no. Geom. Topol. i Teor. Mnozh., 59–73; English transl., Proc. Steklov Inst. Math. **4(247)** (2004), 50–63. MR2168163

[DR] Alexander N. Dranishnikov and Yuli B. Rudyak, *On the Berstein-Svarc theorem in dimension 2*, Math. Proc. Cambridge Philos. Soc. **146** (2009), no. 2, 407–413, DOI 10.1017/S0305004108001904. MR2475974

[FL] F. T. Farrell and J.-F. Lafont, *EZ-structures and topological applications*, Comment. Math. Helv. **80** (2005), no. 1, 103–121, DOI 10.4171/CMH/7. MR2130569

[HS] Bernhard Hanke and Thomas Schick, *The strong Novikov conjecture for low degree cohomology*, Geom. Dedicata **135** (2008), 119–127, DOI 10.1007/s10711-008-9266-9. MR2413333

[K1] T. Kato, *Asymptotic Lipschitz cohomology and higher signatures*, Geom. Funct. Anal. **6** (1996), no. 2, 346–369, DOI 10.1007/BF02247890. MR1384615

[K2] T. Kato, *Asymptotic Lipschitz maps, combable groups and higher signatures*, Geom. Funct. Anal. **10** (2000), no. 1, 51–110, DOI 10.1007/s000390050002. MR1748916

[L] Carl W. Lee, *Subdivisions and triangulations of polytopes*, Handbook of discrete and computational geometry, CRC Press Ser. Discrete Math. Appl., CRC, Boca Raton, FL, 1997, pp. 271–290. MR1730170

[Ma] Varghese Mathai, *The Novikov conjecture for low degree cohomology classes*, Geom. Dedicata **99** (2003), 1–15, DOI 10.1023/A:1024941020306. MR1998926

[M] Ernest Michael, *Continuous selections. I*, Ann. of Math. (2) **63** (1956), 361–382, DOI 10.2307/1969615. MR77107

[MS] John W. Milnor and James D. Stasheff, *Characteristic classes*, Annals of Mathematics Studies, No. 76, Princeton University Press, Princeton, NJ; University of Tokyo Press, Tokyo, 1974. MR440554

[Sc] A. S. Švarc, *The genus of a fibered space* (Russian), Trudy Moskov. Mat. Obšč. **10** (1961), 217–272. MR154284

[Sp] Edwin H. Spanier, *Algebraic topology*, McGraw-Hill Book Co., New York-Toronto-London, 1966. MR210112

[Yu] G. Yu, *The Novikov conjecture* (Russian, with Russian summary), Uspekhi Mat. Nauk **74** (2019), no. 3(447), 167–184, DOI 10.4213/rm9882; English transl., Russian Math. Surveys **74** (2019), no. 3, 525–541. MR3954330

Department of Mathematics, University of Florida, 358 Little Hall, Gainesville, Florida 32611-8105, USA

Email address: dranish@math.ufl.edu

Contemporary Mathematics
Volume **816**, 2025
https://doi.org/10.1090/conm/816/16355

Generalized graph manifolds, residual finiteness, and the Singer conjecture

Luca F. Di Cerbo and Michael Hull

Abstract. We use Price type inequalities to give a geometric proof of the Singer conjecture for extended graph manifolds and pure complex-hyperbolic higher graph manifolds with residually finite fundamental groups. In real dimension three, where a result of Hempel ensures that the fundamental group is always residually finite, we obtain a new proof of a well-known result of Lott and Lück. Finally, we give several classes of higher graph manifolds which do indeed have residually finite fundamental groups.

1. Introduction

In order to state the *Singer conjecture*, we need to define L^2-Betti numbers. L^2-Betti numbers are nonnegative *real* valued numerical invariants associated to a closed Riemannian manifold M and an infinite regular cover $\widetilde{M}$ of such. They were originally introduced by Atiyah in [**Ati76**] in connection with L^2-index theorems. We briefly recall their definition. For an extensive treatment of L^2-Betti numbers and the technical details, we refer to Lück's treatise [**Lüc02**].

Let Γ be a discrete group, and let M be a compact free proper Γ-manifold without boundary endowed with a Γ-invariant Riemannian metric. In other words, one has $M = \widetilde{M}\backslash\Gamma$, where $\Gamma \leq \mathrm{Iso}(\widetilde{M})$ with $\widetilde{M}$ an infinite cover of M, e.g., the *universal* cover or the *universal abelian* cover of M. Define the space of smooth L^2-integrable harmonic k-forms

$$\mathcal{H}^k_{(2)}(\widetilde{M}) \;=\; \left\{\omega \in \Omega^k(\widetilde{M}) \mid \Delta_d\omega = 0, \int_{\widetilde{M}} \omega \wedge *\omega < \infty\right\}$$

where $*$ is the Hodge star operator and $\Delta_d = dd^* + d^*d$ is the Hodge–Laplacian operator. By [**Lüc02**, Section 1.3.2], the spaces $\mathcal{H}^k_{(2)}(\widetilde{M})$ are finitely generated Hilbert modules over the von Neumann algebra $\mathcal{N}(\Gamma)$ of Γ. We define the *L^2-Betti numbers* $b^{(2)}_k(M;\mathcal{N}(\Gamma))$ of (M,Γ) as the von Neumann dimension of the $\mathcal{N}(\Gamma)$-modules $\mathcal{H}^k_{(2)}(\widetilde{M})$:

$$b^{(2)}_k(M;\mathcal{N}(\Gamma)) \;\stackrel{\mathrm{def}}{=}\; \dim_{\mathcal{N}(\Gamma)} \mathcal{H}^k_{(2)}(\widetilde{M}).$$

2020 *Mathematics Subject Classification.* Primary 57N65, 58A10.
Supported in part by NSF grant DMS-2104662.

The L^2-Betti numbers assume values in the extended interval $[0, \infty]$ of the real numbers, and $b_k^{(2)}(M, \mathcal{N}(\Gamma)) \in [0, \infty)$ if the action of Γ is co-compact.

For simplicity's sake, when $\pi\colon \widetilde{M} \to M$ is the topological universal cover of M, we will refer to L^2-Betti numbers as follows:

$$b_k^{(2)}(M) \overset{\text{def}}{=} b_k^{(2)}(M; \mathcal{N}(\Gamma)).$$

As in this paper we are only concerned about L^2-Betti numbers computed with respect to the *universal cover*, we believe this concise notation should not cause any confusion. Finally, we notice that

$$b_k^{(2)}(M) = 0 \iff \mathcal{H}_2^k(\widetilde{M}) = 0; \tag{1}$$

for a proof see Atiyah [**Ati76**] or [**Lüc02**, Chapter I].

We can now state the celebrated and long-standing conjecture of Singer.

CONJECTURE 2 (Singer conjecture). *If M is a closed aspherical manifold of real dimension $2n$, then*

$$b_l^{(2)}(M) = \begin{cases} (-1)^n \chi_{\text{top}}(M) & \text{if} \quad l = n \\ 0 & \text{if} \quad l \neq n. \end{cases}$$

If $\dim_{\mathbb{R}}(M) = 2n + 1$, *we have*

$$b_l^{(2)}(M) = 0, \text{ for any } l.$$

Surprisingly enough, this conjecture is still open even under the assumption that M admits a metric with *strictly negative* sectional curvature! We refer to [**DS22**] for more on the state of the art concerning this problem, and also for some evidence supporting this conjecture under weaker assumptions than negative sectional curvature.

The goal of this paper is to give a geometric proof of the Singer conjecture for two interesting classes of aspherical manifolds: *extended graph manifolds* and *pure complex-hyperbolic higher graph manifolds* with residually finite fundamental group. The class of extended graph manifolds was identified and studied in details by Frigerio, Lafont, and Sisto in [**FLS15**] (specifically, see [**FLS15**, Definition 0.2]). These manifolds are decomposed into finitely many pieces (the vertices), each vertex is a manifold with boundary tori (the edges), and the tori boundary component of various pieces are glued together via affine diffeomorphisms. Moreover, the interior of each vertex is diffeomorphic either to a *complete* finite-volume real-hyperbolic manifold with toral cusps, or to the product of a standard torus with a lower dimensional *complete* finite-volume real-hyperbolic manifold with toral cusps. We denote by $(\mathbb{H}^n_{\mathbb{R}}, g_{\mathbb{R}})$ the real-hyperbolic n-dimensional space with the hyperbolic metric normalized to have sectional curvature equal to -1.

THEOREM 3. *Let M be an extended graph manifold with $k \geq 1$ real-hyperbolic pieces $\{(V_j, g_{\mathbb{R}})\}_{j=1}^k$ and residually finite fundamental group. If* $\dim_{\mathbb{R}}(M) = 2n$, *we have*

$$b_l^{(2)}(M) = \begin{cases} (-1)^n \chi_{\text{top}}(M) = (-1)^n \sum_{j=1}^k \chi_{\text{top}}(V_j) & \text{if} \quad l = n, \\ 0 & \text{if} \quad l \neq n. \end{cases}$$

Finally, if $\dim_{\mathbb{R}}(M) = 2n + 1$ *we have*

$$b_l^{(2)}(M) = 0 \text{ for any } l.$$

The class of higher graph manifolds was first defined and studied by Connell, Suárez, and Serrato in [**CSS19**] (see [**CSS19**, Definition 1]). This class includes as sub-class the extended graph manifolds defined by Frigerio, Lafont, and Sisto, see [**CSS19**, Remark 1]. Roughly speaking, the authors consider graph-like manifolds where the interior of the vertices are diffeomorphic to finite-volume locally symmetric spaces of rank one. Recall that such spaces have finitely many cuspidal ends with nilmanifolds cross-sections. The nilmanifold boundaries are then glued together via affine diffeomorphisms in order to manufacture a closed connected graph-like manifold. Here we are concerned with higher graph manifolds whose vertices are modeled on complex-hyperbolic manifolds with truncated cusps. We refer to such manifolds as pure complex-hyperbolic higher graph manifolds. We denote by $(\mathbb{H}^n_{\mathbb{C}}, g_{\mathbb{C}})$ the complex-hyperbolic n-dimensional space with the hyperbolic metric normalized to have holomorphic sectional curvature equal to -1.

THEOREM 4. *Let M be a pure complex-hyperbolic higher graph manifold with residually finite fundamental group. For $k \geq 2$, let $\{(W_j, g_{\mathbb{C}})\}_{j=1}^k$ denote the complex-hyperbolic pieces. We then have*

$$b_l^{(2)}(M) = \begin{cases} (-1)^n \chi_{\rm top}(M) = (-1)^n \sum_{i=1}^k \chi_{\rm top}(W_j) & \text{if} \quad l = n, \\ 0 & \text{if} \quad l \neq n. \end{cases}$$

REMARK 5. *Because of the finite volume generalization of the Gauss–Bonnet formula (cf. Harder [**Har71**] and Cheeger and Gromov [**CG85**]), we know that for any real-hyperbolic piece $(V_j, g_{\mathbb{R}})$*

$$\chi(V_j) = c(n) Vol_{g_{\mathbb{R}}}(V_j), \tag{6}$$

for some nonzero constant $c(n)$ depending on the dimension only, and where $Vol_{g_{\mathbb{R}}}(V_j)$ is the volume of V_j computed with respect to the normalized real-hyperbolic metric. Similarly, in the complex hyperbolic case we have

$$\chi(W_j) = d(n) Vol_{g_{\mathbb{C}}}(W_j), \tag{7}$$

for some nonzero constant $d(n)$ depending on the dimension only, and where $Vol_{g_{\mathbb{C}}}(W_j)$ is the volume of W_j computed with respect to the normalized complex-hyperbolic metric. By using Equations (6) *and* (7)*, we can alternatively express the L^2-Betti numbers in the middle dimension of Theorem 3 and Theorem 4 in terms of the hyperbolic volume of the complex-hyperbolic pieces.*

There are compelling reasons to believe that Theorem 3 resolves an interesting particular case of the Singer conjecture. Moreover, we argue the general strategy of the proofs we present in this paper may play an important role towards the resolution of this conjecture for nonpositively curved aspherical manifolds. Indeed, as proved by Leeb and Scott in [**LS00**], any nonpositively curved closed Riemannian manifold can be canonically decomposed along finitely many totally geodesically embedded flat closed submanifolds of codimension one. The resulting pieces are either Seifert-fibered or codimension-one atoroidal (see [**LS00**] for the precise definitions). In other words, a general nonpositively curved closed Riemannian manifold has a geometric decomposition very similar to an extended graph manifold. It is then conceivable that by equipping such manifolds with a sequence of Riemannian metrics that marginalizes the Seifert pieces and maximizes the codimension-one atroidal pieces (as we do for extended graph manifolds!), and by studying its pullback to finite covers in conjunction with Price type arguments, one could approach the general case of the Singer conjecture for nonpositively curved Riemannian manifolds. The proof of Theorem 3 nicely outlines this new approach towards the Singer conjecture.

From the perspective of group theory, if M is an extended graph manifold or higher graph manifold then $\pi_1(M)$ splits as a graph of groups where the vertex groups are fundamental groups of the pieces and the edge groups are the fundamental groups of the boundary components of the pieces which are not boundary components of M. Fernós and Valette used a Meyer–Vietoris argument to show if G splits as a graph of groups with infinite edge groups and each vertex group G_v has $b_1^{(2)}(G_v) = 0$, then $b_1^{(2)}(G) = 0$ [**FV17**, Theorem 1]. Together with the vanishing of the first L^2-Betti numbers of the pieces of M (see [**Lüc02**, Chapter 12]), [**FV17**, Theorem 1] implies that $b_1^{(2)}(M) = b_1^{(2)}(\pi_1(M)) = 0$. Using these methods, it is also possible to give algebraic proofs of Theorems 3 and 4.

We note that whenever a group G splits as a graph of groups a formula for $b_1^{(2)}(G)$ in terms of the vertex and edge groups of the splitting is given in [**CHK20**, Theorem 1.5]. Also, if $\pi_1(M)$ satisfies Agol's RFRS condition (a strong form of residual finiteness), then the vanishing of $b_k^{(2)}(\pi_1(M))$ for $1 \leq k \leq n$ is equivalent to the existence of finite index subgroup $\pi_1(M)$ which admits a homomorphism onto $\mathbb{Z}$ with the kernel of type $FP_n(\mathbb{Q})$ [**F21**].

It is unknown whether all extended graph manifolds or all higher graph manifolds have residually finite fundamental group. See [**FLS15**, Section 12.1]. We show that there are many examples which do have residually finite fundamental group, both in the extended graph manifold setting and in the pure complex-hyperbolic higher graph manifold setting.

First, we observe that a result of Huang-Wise [**HW19**] implies the residual finiteness for extended graph manifolds where the pieces are all hyperbolic with *virtually special* fundamental groups in the sense of Wise [**Wis21**]. This virtually special machinery was the key to Agol's solution of the virtual Haken conjecture for 3-manifolds [**Ago13**], and it implies strong separability properties which are well-suited for the construction of finite covers. Moreover, Bergeron, Haglund, and Wise showed that arithmetic real-hyperbolic manifolds have virtually special fundamental group [**BHW11**]. Hence we obtain the following.

THEOREM 8. *Let M be an extended graph manifold with pieces $\{(V_j, g_{\mathbb{R}})\}_{j=1}^k$ such that each V_j is an arithmetic real-hyperbolic manifold. If $\dim_{\mathbb{R}}(M) = 2n$, we have*

$$b_l^{(2)}(M) = \begin{cases} (-1)^n \chi_{\text{top}}(M) = (-1)^n \sum_{j=1}^k \chi_{\text{top}}(V_j) & \text{if} \quad l = n, \\ 0 & \text{if} \quad l \neq n. \end{cases}$$

Finally, if $\dim_{\mathbb{R}}(M) = 2n + 1$ we have

$$b_l^{(2)}(M) = 0, \text{ for any } l.$$

Unfortunately, the virtually special machinery does not apply in the complex-hyperbolic setting. In this setting we use a different approach based on Hempel's proof of the residual finiteness of 3-manifold groups. Using this approach we present a procedure called *iterated (twisted) partial doubles* which, starting with any finite volume real-hyperbolic or complex-hyperbolic manifold with at least two cusps, will construct an infinite family of distinct closed pure real-hyperbolic extended graph manifolds or pure complex-hyperbolic higher graph manifolds respectively with residually finite fundamental group.

Finally, in the last section we show that Theorem 3 can be used to give a new proof of the result of Lück and Lott that the Singer conjecture holds for 3-manifolds. As in [**LL95**], this is proved assuming the geometrization conjecture.

2. Extended graph manifolds and pure complex-hyperbolic higher graph manifolds

In this section, we review the definition and basic properties of extended graph manifolds and higher graph manifolds.

Fix a dimension $n \geq 3$ for the underlying manifold M^n, an integer $k \geq 1$ giving the numbers of geometric pieces, and integers $2 \leq n_i \leq n$ for any $i = 1, \ldots, k$. For every i let V_i be a complete finite-volume real-hyperbolic manifold of dimension n_i with toral cusps. Truncate the cusps of V_i to obtain a compact manifold with torus boundaries say $\bar{V}_i$. For any $i = 1, \ldots, k$ we define the n-dimensional manifold with boundary $N_i := \bar{V}_i \times T^{n-n_i}$, where T^{n-n_i} is the standard $(n-n_i)$-dimensional torus. Notice that the torus boundaries of N_i have a well-defined affine structure.

Definition 9 (cf. [**FLS15**, Definition 0.2]). *For $n \geq 3$ an extended graph n-manifold is a closed manifold built from $k \geq 1$ pieces N_i as above by gluing their torus boundaries by affine diffeomorphisms. An extended graph n-manifold is called* pure real-hyperbolic *if all of the geometric pieces N_i are real-hyperbolic (i.e., $N_i = \bar{V}_i$ for any $i = 1,\ldots,k$).*

It is clear that just like graph 3-manifolds, such manifolds can be nicely described by a graph. Each vertex is one of the geometric pieces V_i, and each edge is the collar of the torus boundary that is obtained by gluing two geometric pieces via an affine diffeomorphism.

Definition 10 ([**Hem87**, Section 2], [**FLS15**, Definitions 2.12 & 2.13]). *Let Γ be a graph with vertex set V and edge set E. Let $\{X_v\}_{v\in V}$ and $\{X_e\}_{e\in E}$ be a collection of topological spaces and let $f_{e,v}\colon X_e \to X_v$ be a π_1-injective embedding whenever v is an endpoint of e. Let X be space constructed by gluing the copy of X_e in X_v to the copy of X_e in X_u via $f_{e,v}^{-1} \circ f_{e,u}$ whenever u and v are the endpoints of e in Γ. Note that we may have $u = v$ but with distinct attaching maps. The space X is a* graph of spaces *with underlying graph Γ. We will often identify the vertex spaces X_v and the edge spaces X_e with their image in X. Without changing the homotopy type of X, we can also assume that each edge space X_e has a neighborhood homeomorphic to $X_e \times (0,1)$.*

When the underlying graph is a tree T, then X is called a tree of spaces. *In this case, in order to keep our terminology consistent with the results in* [**FLS15**], *we will equip X with a map $p\colon X \to T$ which maps each vertex space to the corresponding vertex of T such that for any edge $e \in T$ and t in the internal part e° of e, $X_e \cong p^{-1}(t)$ and $p^{-1}(e^\circ)$ is homeomorphic to $X_e \times (0,1)$. If X is equipped with a Riemannian metric we say that an* internal wall *of X corresponding to $e \in T$ is the closure $p^{-1}(e^\circ)$. Similarly, a* chamber *$C \subset X$ is the pre-image of a vertex of T. Two distinct chambers are said to be* adjacent *if the corresponding vertices in T are joined by an edge. Finally, a wall W is adjacent to the chamber C if $C \cap W \neq \emptyset$.*

Remark 11. *An extended graph n-manifold as in Definition 9 is a graph of spaces where the vertex spaces are the geometric pieces and the edges spaces are* *components of the boundary of these pieces. The universal cover* $\widetilde{M}$ *is a tree of spaces* $(\widetilde{M}, p, T)$, *where for any edge* e *we have that* $p^{-1}(e^\circ)$ *is diffeomorphic to* $\mathbb{R}^{n-1} \times (0,1)$. *For this it is crucial we restrict ourselves to affine diffeomorphism of the boundaries. The main issue here is that one could take connected sums of the edges with exotic spheres; see* [**FLS15**, Remark 2.5] *for more details.*

Next, we observe that $\pi_1(\bar{N}_i)$ injects in the fundamental group of $\pi_1(M)$, where $\pi_1(M)$ is isomorphic to the graph of groups corresponding to the geometric decomposition of M, see [**Hem87**] and [**FLS15**, Section 2.3]. Let $\widetilde{N}_i$ be the universal cover of a piece $\bar{N}_i$. We clearly have

$$\widetilde{N}_i = B_i \times \mathbb{R}^{n-n_i}$$

where B_i is the complement in $\mathbb{H}^{n_i}_{\mathbb{R}}$ of an equivariant family of disjoint open *horoballs.* Putting a rough metric on M as in [**FLS15**] we then get the following result for the topological universal cover $\pi : \widetilde{M} \to M$.

COROLLARY 12. *M admits a Riemannian metric such that $\widetilde{M}$ can be turned into a tree of spaces $(\widetilde{M}, \tilde{p}, \widetilde{T})$ such that:*

- *If C is a chamber of $\widetilde{M}$, then $(C, \tilde{g}_C)$ is isometric to $B \times \mathbb{R}^k$ where the Euclidean factor is equipped with a flat metric, and where B is the complement of open horoballs in $(\mathbb{H}^{n-k}_{\mathbb{R}}, g_{\mathbb{R}})$.*
- *If W is an internal wall of $\widetilde{M}$, then W is diffeomorphic to $\mathbb{R}^{n-1} \times [-1,1]$.*

Corollary 12 tells us that the universal cover of any extended graph manifold is naturally divided into chambers equipped with locally symmetric metrics, either real-hyperbolic or modeled on the product of flat and real-hyperbolic geometries. These chambers are connected via walls diffeomorphic to $\mathbb{R}^{n-1} \times [-1,1]$.

Similar results hold for pure complex-hyperbolic higher graph manifolds, and one can readily adapt the approach given in [**FLS15**, Chapter 2]. The results we state below are somewhat implicit in [**CSS19**].

Fix an integer $n \geq 2$ for the underlying manifold M^{2n}, an integer $k \geq 1$ giving the numbers of complex-hyperbolic pieces of complex dimension n. For every i let W_i be a complete finite-volume complex-hyperbolic manifold of complex dimension n with nilmanifold cusps. In other words, we have that $W_i = \Gamma_i \backslash \mathbb{H}^n_{\mathbb{C}}$ where $\Gamma_i \leq \mathrm{PU}(n,1)$ is a nonuniform torsion-free lattice whose parabolic elements have no rotational part, see [**DD17**, Section 2] for more details. Truncate the cusps of W_i to obtain a compact manifold with nilmanifold boundaries, say $\bar{W}_i$. For any $i = 1, \ldots, k$ we define the $2n$-dimensional manifold with boundary $M_i := \bar{W}_i$. Notice that the nilmanifold boundaries of M_i have a well-defined affine structure, see [**Gol99**, Sections 4.2.1 and 4.2.2]. Indeed, each horosphere in $\mathbb{H}^n_{\mathbb{C}}$ has a geometric structure modeled on the $(2n-1)$-Heisenberg group $\mathcal{H}^{2n-1}$ and the nilmanifold cross sections of the cusps are simply quotients of $\mathcal{H}^{2n-1}$ via lattices of translations.

DEFINITION 13 (cf. [**CSS19**, Definition 1]). *For $n \geq 2$ a pure complex-hyperbolic higher graph $2n$-manifold is a closed manifold built from $k \geq 1$ pieces M_i as above by gluing their nilmanifold boundaries with affine diffeomorphisms.*

Similar to the case of extended graph manifolds, we observe that $\pi_1(W_i) = \pi_1(\bar{M}_i)$ injects in the fundamental group of $\pi_1(M)$, where $\pi_1(M)$ is isomorphic to the graph of groups corresponding to the geometric decomposition of M; see again

[**Hem87**] and [**FLS15**, Section 2.3]. Let $\widetilde{M}_i$ be the universal cover of a piece $\bar{M}_i$. We clearly have

$$\widetilde{M}_i = B_i$$

where B_i is the complement in $\mathbb{H}^n_{\mathbb{C}}$ of an equivariant family of disjoint open *horoballs*. Putting a rough metric on M as in Corollary 12, we then get the following result for the topological universal cover $\pi : \widetilde{M} \to M$.

COROLLARY 14. *M admits a Riemannian metric such that $\widetilde{M}$ can be turned into a tree of spaces $(\widetilde{M}, \tilde{p}, \widetilde{T})$ such that:*

- *If C is a chamber of $\widetilde{M}$, then $(C, \tilde{g}_C)$ is isometric to B, where B is the complement of open horoballs in $(\mathbb{H}^n_{\mathbb{C}}, g_{\mathbb{C}})$.*
- *If W is an internal wall of $\widetilde{M}$, then W is diffeomorphic to $\mathcal{H}^{2n-1} \times [-1, 1]$, where $\mathcal{H}^{2n-1}$ is the $(2n-1)$-dimensional Heisenberg Lie group.*

3. Dimension estimates

In this section, we repackage the Price inequality estimates for harmonic forms proved by L. F. Di Cerbo and M. Stern in [**DS22**], [**DS19**], and [**DS20**]. The statements presented here are tailored to the study of higher graph manifolds, and they may slightly differ from the presentation in the original sources.

Let (M^n, g) be a closed Riemannian manifold, and denote by $\mathcal{H}^k_g(M)$ the finite dimensional vector space of harmonic k-forms. By the Hodge–de Rham theorem, we know that

$$b_k(M) = \dim_{\mathbb{R}} \mathcal{H}^k_g(M)$$

where $b_k(M)$ is the kth topological Betti number of M.

Let $K(\cdot, \cdot)$ denote the Schwartz kernel for the L^2 orthogonal projection onto $\mathcal{H}^k_g(M)$. Thus for $x, y \in M$, we have

$$K(x, y) \in Hom(\Omega^k T^*_y M, \Omega^k T^*_x M).$$

Given an L^2-orthonormal basis $\{\alpha_j\}_{j=1}^l$ for $\mathcal{H}^k_g(M)$, we have

$$K(x, y) = \sum_{j=1}^{l} \alpha_j(x) \langle \cdot, \alpha_j(y) \rangle, \tag{15}$$

so that

$$TrK(x, x) = \sum_{i=1}^{\binom{n}{k}} \langle K(x, x)(e_i), e_i \rangle = \sum_{i=1}^{b_k(M)} |\alpha_i|^2_x \geq 0, \tag{16}$$

and

$$\int_M TrK(x, x) d\mu_g = b_k(M). \tag{17}$$

Because of Equation (17), we give the following definition.

DEFINITIONS 18. *Given (M, g), $\mathcal{H}^k_g(M)$, and an L^2-orthonormal basis*

$$\{\alpha_j\}_{j=1}^{b_k(M)}$$

for $\mathcal{H}^k_g(M)$, we define the pointwise kth Betti number density to be

$$\rho_k(x) := TrK(x, x),$$

for any $x \in M$.

Next, we collect a linear algebra type pointwise estimate for the kth Betti number density.

LEMMA 19. *Given* (M, g) *and* $\mathcal{H}^k_g(M)$ *as above, for any* $x \in M$ *we have*

$$0 \leq \rho_k(x) \leq \binom{n}{k} \sup_{\alpha \in \mathcal{H}^k_g(M): \|\alpha\|^2_{L^2}=1} |\alpha(x)|^2.$$

PROOF. See [**DS19**, Lemma 14]. □

The following lemma is an application of the usual elliptic regularity and Moser iteration for harmonic forms in bounded geometry.

LEMMA 20. *Given* (M^n, g) *and* $\mathcal{H}^k_g(M)$ *as above, if the sectional curvature is bounded*

$$|sec_g| \leq a$$

for some $a \geq 0$, *then there exists a constant* $c = c(n, a, k, \mathrm{inj}_g(M, x)) > 0$ *such that for any* $x \in M$

$$0 \leq \rho_k(x) \leq c,$$

where $\mathrm{inj}_g(M, x)$ *is the injectivity radius at* x.

PROOF. See [**DS19**, Lemma 16]. □

Finally, we collect a couple of estimates for the pointwise Betti number density at points in closed Riemannian manifolds which possess geodesic balls centered around them isometric to large geodesic balls in the real– and complex-hyperbolic space. We start with the real-hyperbolic case.

LEMMA 21. *Let* (M^n, g) *and* $\mathcal{H}^k_g(M)$ *be as above. For any* $R > 1$ *and* $x \in M$ *such that* $B_g(x, R)$ *is isometric to a ball* $B_{g_{\mathbb{R}}}(-, R)$ *in the real-hyperbolic space form* $(\mathbb{H}^n_{\mathbb{R}}, g_{\mathbb{R}})$ *(with sectional curvature normalized so that* $sec_{g_{\mathbb{R}}} = -1$*), we have*

$$0 \leq \rho_k(x) \leq \begin{cases} C_1(n,k)e^{-(n-1-2k)R} & \text{if} \quad n-1-2k > 0, \\ C_2(n,k)R^{-1} & \text{if} \quad n-1-2k = 0, \end{cases} \tag{22}$$

where $C_i(n, k) > 0$, $i = 1, 2$, *depend on* n *and* k *only.*

PROOF. The lemma is a reformulation, tailored to our present purposes, of the estimates for negatively ϵ-pinched Riemannian manifolds originally proved in [**DS22**, Theorems 87, 96, and Corollary 108]. Set $\epsilon = 0$ in these estimates to extract the results needed for the proof of this lemma. □

In the complex-hyperbolic case we have the following.

LEMMA 23. *Let* (M^{2n}, g) *and* $\mathcal{H}^k_g(M)$ *be as above. For any* $R > 1$ *and* $x \in M$ *such that* $B_g(x, R)$ *is isometric to a ball* $B_{g_{\mathbb{C}}}(-, R)$ *in the complex-hyperbolic space form* $(\mathbb{H}^n_{\mathbb{C}}, g_{\mathbb{C}})$ *(with sectional curvature normalized so that* $-4 \leq sec_{g_{\mathbb{C}}} \leq -1$*), we have*

$$0 \leq \rho_k(x) \leq \begin{cases} D_1(n,k)e^{-2(n-1-k)R} & \text{if} \quad n-1-k > 0, \\ D_2(n,k)R^{-1} & \text{if} \quad n-1-k = 0, \end{cases} \tag{24}$$

where $D_i(n, k) > 0$, $i = 1, 2$ *depends on the* n *and* k *only.*

PROOF. For this lemma, we need to go beyond the estimates proved in [**DS22**] for negatively curved and pinched Riemannian manifolds. Indeed, the estimate in (24) crucially relies upon the fact that geodesic spheres in complex-hyperbolic space have constant mean curvature. For the details see [**DS20**, Section 3] and, in particular, Lemma 25 and Lemma 28 therein. □

4. Cheeger–Gromov–Fukaya theory and Price inequalities: Proofs of Theorems 3 and 4

In the paper [**CSS19**], Connell, Suárez, and Serrato compute exactly the minimal volume of extended graph manifolds (see [**CSS19**, Theorem 4]), and among other things are able to estimate from above and below the minimal volume of *any* higher graph manifolds (see [**CSS19**, Theorem 3]). An important ingredient in their analysis is the construction of a particular sequence of metrics on any given higher graph manifold. This sequence of metrics is identically equal to the locally symmetric hyperbolic metrics on larger and larger portions of the pure hyperbolic pieces, and it collapses with bounded curvature the nonpure pieces and the gluing regions. The construction of such sequence of metrics relies on the general theory of collapsing developed by Cheeger-Gromov [**CG86**] and a crucial subsequent refinement of Fukaya [**Fuk89**]. The following lemma is proven in [**CSS19**, Section 2, pp. 1287–1292]. We prefer to collect as a lemma a particular case of the construction of Connell, Suárez, and Serrato when specialized to extended graph manifolds (as defined in [**FLS15**]). For a very detailed and elementary proof of this lemma in the 3-dimensional case, we also refer the interested reader to the Ph.D. thesis of Pieroni [**Pie19**].

LEMMA 25 (Connell, Suárez, and Serrato). *Let M be an extended graph manifold with $k \geq 1$ real-hyperbolic pieces $\{(V_j, g_{\mathbb{R}})\}_{j=1}^k$. There exists a sequence of metrics $\{g_m\}$ on M such that:*

- $\lim_{m\to\infty} Vol_{g_m}(M) = \sum_{j=1}^k Vol_{g_{\mathbb{R}}}(V_j)$;
- $|sec_{g_m}| \leq 1 + \delta_m$, *where* $\lim_{m\to\infty} \delta_m = 0$;
- $g_m = g_{\mathbb{R}}$ *on each geometric piece* V_j *up to height* m *on each of its cusps. In particular,* $\lim_{m\to\infty} diam_{g_m}(M) = \infty$.

PROOF. Recall that, if $n = \dim M$, a cusp of a real-hyperbolic piece is isometric to $\mathbb{R}^+ \times T^{n-1}$ equipped with the metric $dt^2 + e^{-2t} g_{T^{n-1}}$, where $g_{T^{n-1}}$ is a flat metric on a $(n-1)$-dimensional torus T^{n-1}. Up to height m on a cusp then corresponds the region $[0, m] \times T^{n-1}$. For the remaining details of the construction of the sequence of metrics g_m, we refer to [**CSS19**, Section 2]. □

We can now prove the first theorem stated in the introduction.

PROOF OF THEOREM 3. Let $\Lambda := \pi_1(M)$ be residually finite, and let $\{\Lambda_i\}$ be a cofinal filtration of Λ by nested finite index normal subgroups. For any fixed integer m, let (M^i, g_m^i) be the associated tower of normal Riemannian coverings where

$$\pi_i : M^i \to M, \quad g_m^i = \pi_i^*(g_m)$$

are, respectively, the cover associated to the finite index normal subgroup $\Lambda_i \triangleleft \Lambda$ and the pull-back metric of g_m via π_i. Here (M, g_m) is the sequence of metrics given in Lemma 25.

For any integer m, we define M_m to be the possibly disconnected open submanifold of M given by the union up to height $m/2$ along the cusps of all the pure pieces $\{V_j\}_{j=1}^k$. Also, for any regular cover $\pi_i : M^i \to M$ we define

$$M_m^i := \pi_m^{-1}(M_m) \subset M^i,$$

where M_m^i is an open and possibly disconnected submanifold of M^i.

LEMMA 26. *For any integer $m \geq 1$, there exists a positive integer i_m such that for $i \geq i_m$ and for $x \in M_m^i$ there exists a ball centered at x of radius $m/2$, say $B_{g_i}(x, m/2)$, isometric to a ball in $(\mathbb{H}_{\mathbb{R}}^n, g_{\mathbb{R}})$.*

PROOF. First, for any $i, m \geq 1$ define

$$r_m^i \stackrel{\text{def}}{=} \inf\{d_{\tilde{g}_m}(z, \lambda_i z) \mid z \in \widetilde{M}, \quad \lambda_i \in \Lambda_i, \quad \lambda_i \neq \mathrm{id}_{\Lambda_i}\}$$

where $\tilde{\pi} : \widetilde{M} \to M$ is the topological universal cover and $\tilde{g}_m := \tilde{\pi}^* g_m$. By [**DL19**, Theorem 3.3], we know that for any fixed m

$$\lim_{i\to\infty} r_m^i = \infty. \tag{27}$$

Moreover, a Riemannian ball $B_{g_m^i}(-, r_m^i/2)$ (centered at any point) in (M^i, g_m^i) is isometric to a ball of the same radius in $(\widetilde{M}, \tilde{g}_m)$. By construction, a ball $B_{g_m^i}(x, m/2)$ for $x \in M_m^i$ contains only points where the sectional curvature is equal to -1. Next, choose i_m so that for $i \geq i_m$ we have $r_m^i \geq m/2$. This can always be achieved thanks to Equation (27). We then have that for $i \geq i_m$, the ball $B_{g_m^i}(x, m/2)$ for $x \in M_m^i$ must be entirely contained in a real-hyperbolic chamber $(C, g_{\mathbb{R}})$ of the tree of spaces $(\widetilde{M}, \tilde{\pi}, \widetilde{T})$; see Corollary 12. □

Using Equation (17) and Definition 18, given any closed n-dimensional Riemannian manifold (M, g) one can write

$$b_l(M) = \int_M \rho_l(x) d\mu_g$$

where $\rho_l(x)$ is the lth Betti number density, see Section 3 for more details. Consider first the Betti numbers of subcritical degree $l < (\dim_{\mathbb{R}}(M) - 1)/2$. We then have the estimate

$$b_l(M^i) = \int_{M_m^i} \rho_l(x) d\mu_{g_m^i} + \int_{M^i \setminus M_m^i} \rho_l(x) d\mu_{g_m^i}. \tag{28}$$

By using Lemma 26, we know that if $i > i_m$ then any point $x \in M_m^i$ has a ball $B_{g_i}(x, m/2)$ isometric to a ball in the real-hyperbolic space $(\mathbb{H}_{\mathbb{R}}^n, g_{\mathbb{R}})$. Moreover, by eventually increasing i_m, we can also arrange that if $i > i_m$ then any point $x \in M^i \setminus M_m^i$ satisfies

$$inj_{g_m^i}(M^i, x) \geq \delta, \tag{29}$$

for some $\delta > 0$. To see this, first recall that by [**FLS15**, Lemma 3.2] any extended graph manifold is a $K(\pi, 1)$-space and then an *essential* manifold in the terminology of Cheeger and Gromov [**CG86**]. Now by construction of M (as a graph), we have that M admits an F-structure outside a compact subset, say C, compatible with the sequence of metrics $\{g_m\}$. The complement of C in M clearly contains all the nonpure pieces and the near infinity part of the pure real-hyperbolic pieces, i.e., the near infinity parts of the toral cusps. By [**CG86**, Theorem 10] the F-structure is injective, and as observed in [**Ron93**, Theorem 2.5], this easily implies that the sequence of metrics $\{g_m\}$ is collapsing with *bounded covering geometry.* As the sequence (M^i, g^i_m) is converging to the universal Riemannian cover, this is the same as saying that Equation 29 holds (with δ independent of m) for i large enough depending on m.

Next by Lemma 21, if $x \in M^i_m$ and $i > i_m$ we have

$$0 \leq \rho_l(x) \leq C_1(\dim_{\mathbb{R}}(M), l)e^{-(\dim_{\mathbb{R}}(M)-1-2l)\frac{m}{2}}, \tag{30}$$

where $C_1(\dim_{\mathbb{R}}(M), l) > 0$ is a constant depending on the dimension and the degree only. Similarly, by Lemma 20 if $x \in M^i \setminus M^i_m$ and $i > i_m$

$$0 \leq \rho_l(x) \leq c(\dim_{\mathbb{R}}(M), l, 1+\delta_m, \delta) \tag{31}$$

where δ is the injectivity radius lower bound coming from collapsing with bounded covering geometry given in Equation (29), and where $1+\delta_m$ is the sectional curvature bound for the sequence of metrics $\{g_m\}$ given in Lemma 25. As $\lim_{m\to\infty} \delta_m = 0$ (see Lemma 25), for $m >> 0$ we can always assume that $0 < 1+\delta_m \leq 2$. Thus, for large m the upper bound in Equation (31) can be chosen to be independent of m.

Next, observe that by construction of the metrics $\{g_m\}$ on M we have

$$\lim_{m\to\infty} Vol_{g_m}(M \setminus M_m) = 0. \tag{32}$$

By combining Equations (28), (30), (31), and (32) we have the estimate

$$\begin{aligned} b_l(M^i) &\leq C_1 Vol_{g^i_m}(M^i_m)e^{-(\dim_{\mathbb{R}}(M)-1-2l)\frac{m}{2}} + cVol(M^i \setminus M^i_m) \\ &\leq C_1 Vol_{g^i_m}(M^i)e^{-(\dim_{\mathbb{R}}(M)-1-2l)\frac{m}{2}} + c\deg(\pi_i)\epsilon(m), \end{aligned}$$

where $\epsilon(m) > 0$ is such that

$$\lim_{m\to\infty} \epsilon(m) = 0,$$

see Equation (32). We then compute

$$\limsup_{i\to\infty} \frac{b_l(M^i)}{Vol_{g^i_m}(M^i)} < C_1 e^{-(\dim_{\mathbb{R}}(M)-1-2l)\frac{m}{2}} + \frac{c\cdot\epsilon(m)}{Vol_{g_m}(M)}.$$

Since the constant $c(n, l, 1+\delta_m, \delta)$ can be chosen independently of m for $m >> 0$, there exists $m_0 > 0$ such that for any $m \geq m_0$ we have

$$\limsup_{i\to\infty} \frac{b_l(M^i)}{\deg(\pi_i)} \leq C_1 Vol_{g_m}(M)e^{-(\dim_{\mathbb{R}}(M)-1-2l)\frac{m}{2}} + c\cdot\epsilon(m), \tag{33}$$

where $c = c(\dim_{\mathbb{R}}(M), l, \delta) > 0$. By Lück's approximation theorem (see [**Lüc94**, Main Theorem]), we know that

$$\limsup_{i\to\infty} \frac{b_l(M^i)}{\deg(\pi_i)} = \lim_{i\to\infty} \frac{b_l(M^i)}{\deg(\pi_i)} = b^{(2)}_l(M). \tag{34}$$

Thus, by taking the limit as $m \to \infty$ in the right-hand side of Equation (33) and combining it with Equation (34), we obtain that $b^{(2)}_l(M) = 0$ for $l < (\dim_{\mathbb{R}}(M) - 1)/2$, and by Poincaré duality for $l > (\dim_{\mathbb{R}}(M) - 1)/2$. Finally, if $\dim_{\mathbb{R}}(M) = 2n$ and $l = n$, we can use these vanishing results to compute

$$b^{(2)}_n(M) = \lim_{i\to\infty} \frac{b_n(M^i)}{\deg(\pi_i)} = \frac{(-1)^n\chi(M^i)}{\deg(\pi_i)} = (-1)^n\chi_{top}(M).$$

Next, a Meyer–Vietoris argument yields

$$\chi_{top}(M) = \sum_{j=1}^{k} \chi(V_j),$$

and this concludes the proof in the even dimensional case.

In the odd dimensional case, everything is the same with the exception of the critical degree $l = (\dim_{\mathbb{R}}(M) - 1)/2$. In this degree, Lemma 21 implies that if $x \in M^i_m$ and $i > i_m$ we have

$$0 \leq \rho_l(x) \leq C_2(\dim_{\mathbb{R}}(M), l)\frac{2}{m}, \tag{35}$$

cf. Equation (30). The rest of the argument now goes through simply interchanging Equation (30) with Equation (35). As the upper bound in Equation (35) goes to zero when $m \to \infty$, we still obtain in the end that $b^{(2)}_l(M) = 0$ and then by Poincaré duality $b^{(2)}_{l+1}(M) = 0$ also. The proof is complete. □

We can now focus on the case of pure complex-hyperbolic higher graph manifolds. The following lemma provides a suitable sequence of metrics on such manifolds which collapses with bounded covering geometry the gluing regions in the graph manifold, and it is locally symmetric on larger and larger portions of the vertices. This sequence of metrics was explicitly constructed by Connell, Suárez, and Serrato in [**CSS19**].

LEMMA 36 (Connell, Suárez, and Serrato). *Let M^{2n} be a pure complex-hyperbolic higher graph manifold with $k \geq 2$ pure complex-hyperbolic pieces, say $\{(W_j, g_{\mathbb{C}})\}_{j=1}^k$. There exists a sequence of metrics $\{g_m\}$ on M such that:*

- $\lim_{m\to\infty} Vol_{g_m}(M) = \sum_{j=1}^k Vol_{g_{\mathbb{C}}}(W_j)$*;*
- $|sec_{g_m}| \leq 4$ *for any* $m \geq 1$*;*
- $g_m = g_{\mathbb{C}}$ *on each piece* W_j *up to height* m *on each of its cusps. In particular,* $\lim_{m\to\infty} diam_{g_m}(M) = \infty$.

PROOF. See [**CSS19**, Section 2, pp. 1287–1293]. □

We can now prove the second theorem stated in the introduction.

PROOF OF THEOREM 4. The set up is very similar to the proof of Theorem 3. Let $\Lambda := \pi_1(M)$ be residually finite, and let $\{\Lambda_i\}$ be a cofinal filtration of Λ by nested finite index normal subgroups. For any fixed integer m, let (M^i, g^i_m) be the associated tower of normal Riemannian coverings where

$$\pi_i : M^i \to M, \quad g^i_m = \pi_i^*(g_m)$$

are, respectively, the cover associated to the finite index normal subgroup $\Lambda_i \lhd \Lambda$ and the pull-back metric of g_m via π_i. Here (M, g_m) is the sequence of metrics given in Lemma 25.

For any integer m, we define M_m to be the possibly disconnected open submanifold of M given by the union up to height $m/2$ along the cusps of all the pure pieces $\{V_j\}_{j=1}^k$. Also, for any regular cover $\pi_i : M^i \to M$ we define

$$M^i_m := \pi_m^{-1}(M_m) \subset M^i$$

where M^i_m is an open and possibly disconnected submanifold of M^i.

LEMMA 37. *For any integer $m \geq 1$, there exists a positive integer i_m such that for $i \geq i_m$ and for $x \in M^i_m$ there exists a ball centered at x of radius $m/2$, say $B_{g_i}(x, m/2)$, isometric to a ball in $(\mathbb{H}^n_{\mathbb{C}}, g_{\mathbb{C}})$.*

PROOF. Use Corollary 14 instead of Corollary 12 in the proof of Lemma 26. The rest is identical. □

Given Lemma 26 and Lemma 36, we write for $l < n = \dim_{\mathbb{R}}(M)/2$

$$b_l(M) = \int_M \rho_l(x) d\mu_g$$

where $\rho_l(x)$ is the lth Betti number density, see Section 3 for more details. From this point on, one can then follow step-by-step the proof of Theorem 3 in the even dimensional case. The only difference is the use of the Price inequality estimate for complex-hyperbolic spaces given Lemma 23 instead of the estimate for real-hyperbolic spaces given in Lemma 21. The proof is complete. □

5. Higher graph manifolds without pure pieces

Recall the definition of Gromov minimal volume.

DEFINITION 38 (Gromov [**Gro82**]). *Let M^n be a smooth manifold. Consider the set of all complete metrics on M satisfying the bound*

$$|sec_g| \leq 1.$$

We define the Gromov minimal volume of M to be

$$\mathrm{MinVol}(M) := \inf_g \left\{ Vol_g(M) \mid |sec_g| \leq 1 \right\}.$$

As stated in [**Gro99**], for each n there is a constant c_n with the following property: If M^n is an n-dimensional, closed, aspherical manifold, then

$$b^{(2)}_i(M) \leq c_n MinVol(M), \tag{39}$$

for all $i \geq 0$. In [**Gro99**, Section 5.33] Gromov outlines a proof of this fact under the assumption that $\pi_1(M)$ is residually finite. See the paper by Sauer [**Sau09**] for a proof of Equation (39) without the residual finiteness assumption. In particular, the Singer conjecture holds true for aspherical manifolds with $MinVol = 0$ (see Conjecture 2). Finally, we observe that if the fundamental group is residually finite then Fauser, Friedl, and Löh prove an integral approximation theorem for the simplicial volume of higher graph manifolds without pure pieces, see [**FFL19**, Theorem 1.2]. This interesting result can then be used to prove the vanishing of gradient invariants of higher graph manifolds without pure pieces, see [**FFL19**, Corollary 1.4 and Remark 1.5]. Gradient invariants are defined in [**FFL19**] as the normalization over finite coverings and finite index subgroups of several topological quantities. In the case of Betti numbers, this normalization process recovers the usual L^2-Betti numbers. For more details, see [**FFL19**, Introduction].

We now provide a proof of the Singer conjecture (and Equation (39)) for aspherical manifold with $MinVol = 0$ and residually finite fundamental group. The proof is along the lines of the strategy employed in the proofs of Theorem 3 and Theorem 4. Even if this was already known thanks to the work of Gromov and Sauer, we think it is somewhat interesting to have yet another approach.

PROPOSITION 40. *Let M^n be a closed aspherical manifold with $MinVol(M)=0$ and $\pi_1(M)$ residually finite. We then have*

$$b_l^{(2)}(M) = 0, \quad for \quad l = 0, 1, \dots, n.$$

PROOF. Since $MinVol(M) = 0$, there exists a sequence of metrics $\{g_m\}$ on M, with $|sec_{g_m}| \leq 1$, and such that

$$\lim_{m \to \infty} Vol_{g_m}(M) = 0.$$

Given $\Lambda := \pi_1(M)$, let a $\{\Lambda_i\}$ be a cofinal filtration by nested finite index normal subgroups. For any fixed m, let (M^i, g^i_m) be the associated tower of normal Riemannian coverings where

$$\pi_i : M^i \to M, \quad g^i_m = \pi_i^*(g_m)$$

are, respectively, the cover associated to $\Lambda_i \triangleleft \pi_1(M)$ and the pull-back metric of g_m via π_i. Next, we have the following lemma.

LEMMA 41. *For any fixed m, there exists $i_m > 0$ such that for any $i \geq i_m$ we have*

$$\text{inj}_{g^i_m}(M^i) \geq \epsilon > 0, \tag{42}$$

where $\epsilon > 0$ is independent of m.

PROOF. M is an essential manifold in the terminology of Cheeger and Gromov [**CG86**], and we have that M admits an F-structure. By [**CG86**, Theorem A.1] the F-structure is injective, and as observed in [**Ron93**, Theorem 2.5] this easily implies that the sequence of metrics $\{g_m\}$ is collapsing with *bounded covering geometry.* As the sequence (M^i, g^i_m) is converging to the universal Riemannian cover, this gives that Equation (42) is satisfied. □

We can now use Lemma 41 with Lemma 20, to conclude that

$$\frac{b_l(M^i)}{Vol_{g^i_k}(M^i)} \leq C, \tag{43}$$

for $i \geq i_m$, with $C > 0$ independent of m. We can now apply Lück's approximation theorem (see Main Theorem in [**Lüc94**]) to conclude that for any fixed m

$$\lim_{i \to \infty} \frac{b_l(M^i)}{Vol_{g^i_m}(M^i)} = \lim_{i \to \infty} \frac{b_l(M^i)}{\deg(\pi_i) Vol_{g_m}(M)} = \frac{b_l^{(2)}(M)}{Vol_{g_m}(M)}. \tag{44}$$

By combining Equations (43) and (44), we then obtain that

$$b_l^{(2)}(M) \leq C \cdot Vol_{g_m}(M),$$

for any m (*cf.* Equation (39)). The proof is therefore complete. □

The next corollary is then clear.

COROLLARY 45. *Let M^n be a higher graph manifold without pure pieces and with $\pi_1(M)$ residually finite. We then have*

$$b_l^{(2)}(M) = 0$$

for all $l \geq 0$.

PROOF. By [**CSS19**, Theorem 1], we know that $MinVol(M) = 0$. By [**FLS15**, Lemma 3.2], we know that M is aspherical. Conclude the proof by using Proposition 40. □

6. Residual finiteness of higher graph manifolds

In this section, we study the residual finiteness of fundamental groups of higher graph manifolds. In general, it is an open question whether all such manifolds have residually finite fundamental group; see [**FLS15**, Section 12.1]. We will first show that a large class of pure real-hyperbolic extended graph manifolds have residually finite fundamental group using Wise's virtually special machinery (Theorem 46). This class includes, for example, any such manifold where the pieces are all arithmetic (Corollary 47). However, this machinery does not apply in the complex-hyperbolic setting. Hence, we will study applications of Hempel's approach to the residual finiteness of 3-manifold groups in higher dimensions. A conjecture of Behrstock and Neumann implies that this approach works quite broadly; see Proposition 54. However, even without this conjecture we are able to apply this approach to a procedure we call *iterated (twisted) partial doubles* to construct infinite families of both pure real-hyperbolic extended graph manifolds and pure complex-hyperbolic higher graph manifolds. Finally, we give some explicit examples of manifolds which can be used to construct these families. The pure complex-hyperbolic higher graph manifolds we construct appear to be the *first* explicit examples of such manifolds available in the literature.

6.1. Wise's virtually special machinery and pure arithmetic real-hyperbolic higher graph manifolds. First, we observe that Wise's virtually special machinery implies residual finiteness of the fundamental groups of a large class of pure real-hyperbolic extended graph manifolds. This general result applies in particular to the case where all pieces are real-hyperbolic and arithmetic, though we expect it to apply more broadly as well. In fact, we are not aware of any real-hyperbolic manifolds which are known to not satisfy the hypothesis of Theorem 46.

A cube complex is a space built by gluing cubes of various dimensions together along faces, where a cube is a copy of $[0,1]^n$ for some n. A cube complex C is called *special* if C admits a local isometry into the Salvetti complex of a right-angled Artin group, and C is *virtually special* if it has a finite cover which is special. A group G is called virtually special if $G \cong \pi_1(C)$ where C is a virtually special cube complex. This implies that G has a finite index subgroup which embeds into a right-angled Artin group. G is *virtually compact special* if, in addition, C is compact and *virtually sparse special* if C is the union of a compact subset and certain cusp-like subspaces. We refer to [**Wis21**] for precise definitions and properties of virtually special groups.

The next result is a special case of [**HW19**, Theorem 1.9]. To see that the hypotheses are satisfied, we note that by Dahmani's combination theorem for relatively hyperbolic groups [**Dah03**] $\pi_1(M)$ will be hyperbolic relative to the collection of maximal parabolic subgroups of the fundamental group of each piece. These parabolic subgroups will be relatively quasi-convex by definition [**Hru10**].

THEOREM 46. *Let M be a pure real-hyperbolic extended graph manifold such that the fundamental group of each piece is virtually sparse special. Then $\pi_1(M)$ is residually finite.*

Arithmetic lattices were shown to be virtually special in [**BHW11**]. While they focused on uniform lattices, the construction produces a virtually special but not compact cube complex for nonuniform lattices as well, see [**BHW11**, Section 9] and [**BW12**, Section 4]. Finiteness properties of this method of producing cube complexes were studied in [**HW14**], where they were shown to be relatively compact in the nonuniform case, see [**HW14**, Corollary 7.15]. They also show that the additional condition of the stronger property of being sparse also holds provided one makes an appropriate choice of cubulation of the peripheral structure, see [**HW14**, Theorem 7.12].

COROLLARY 47. *Let M be a pure real-hyperbolic extended graph manifold such that each piece is a real-hyperbolic arithmetic manifold. Then $\pi_1(M)$ is residually finite.*

Corollary 47 together with Theorem 3 implies Theorem 8.

6.2. Hempel's criterion and the cusp covering conjecture. The virtually special machinery cannot be applied in the complex-hyperbolic setting, see [**BW12**, Remark 4.4]. In order to produce examples of pure complex-hyperbolic higher graph manifolds with residually finite fundamental group, we extend Hempel's approach to residual finiteness for 3-manifold groups to certain classes of higher graph manifolds.

Hempel's proof has two main ingredients, separability and compatibility. Compatibility is a condition which allows a collection of covers of the geometric pieces to be assembled into a cover of the whole manifold. When combined with separability of the boundary subgroups, any loop in the manifold can be lifted to a path in the finite cover assembled in this way, which shows that the fundamental group is residually finite. We now describe these conditions in more detail.

Let X be a graph of spaces (see Definition 10), and let $V_1, \dots, V_m$ denote the vertex spaces of X. For each V_i, let $\overline{V}_i$ denote a covering space of V_i with covering map $p_i \colon \overline{V}_i \to V_i$. The collection of covers $\overline{V}_1, \dots, \overline{V}_m$ is called *compatible* if, whenever V_i and V_j are adjacent vertex spaces glued along an edge space E, then for each component $\overline{E}$ of $p_i^{-1}(E)$ and each component $\overline{E}'$ of $p_j^{-1}(E)$, the induced covers $\overline{E} \to E$ and $\overline{E}' \to E$ are equivalent. Note that this implies in particular that for any two components of $p_i^{-1}(E)$ the covering maps to E must be equivalent.

Any compatible collection of finite covers can be assembled into a finite cover of X; see [**Hem87**, Theorem 2.2].

THEOREM 48. *Let X be a graph of spaces and let $V_1, \dots, V_m$ denote the vertex spaces of X. Let $\overline{V}_1, \dots, \overline{V}_m$ be a compatible collection of finite covers. Then X has a finite cover $\overline{X}$ which is a graph of spaces where each vertex space is a copy of some $\overline{V}_i$.*

By the Seifert–Van Kampen theorem $\pi_1(X)$ splits as graph of groups with vertex groups given by fundamental groups of the vertex spaces and edge groups given by fundamental groups of the edge spaces; see [**FLS15**, Section 2.3]. Dual to this graph of groups decomposition of $\pi_1(X)$ is an action of $\pi_1(X)$ on a tree, called the Bass–Serre tree of the splitting. In this case the Bass–Serre tree can be identified with the tree T coming from the tree of spaces structure on the universal cover $\widetilde{X}$, with the action of $\pi_1(X)$ on T induced by the action of $\pi_1(X)$ on $\widetilde{X}$. Each vertex of T has a stabilizer which is conjugate to a vertex group in the graph of groups and each adjacent edge stabilizer is conjugate to an adjacent edge group.

Next, we give an algebraic version of the above compatibility condition in terms of this graph of groups structure. Suppose G is a group which splits as a graph of groups with vertex set V and edge set E and let $\{G_v\}_{v\in V}$ and $\{G_e\}_{e\in E}$ denote the corresponding vertex and edge groups in this graph of groups. The graph of groups structure comes with injective homomorphisms $\phi_{e,v}\colon G_e \to G_v$ whenever v is an endpoint of the edge e. However, we will identify G_e with the subgroup $\phi_{e,v}(G_e)$ of G_v and drop the maps $\phi_{e,v}$ from the notation. Now collections of finite index normal subgroups $\{N_v \le G_v\}_{v\in V}$ and $\{N_e \le G_e\}_{e\in E}$ are called *compatible* if whenever v is an endpoint of the edge e, then $N_v \cap G_e = N_e$. In the case where $G = \pi_1(X)$ for a graph of spaces X and the graph of groups is the one described above, then given a compatible collection of finite index normal subgroups, the covers induced by these subgroups will be a compatible collection of covers.

The second ingredient in Hempel's approach is the *separability* of the boundary subgroups of each geometric piece. This algebraic condition ensures that a loop based at a point on the boundary can be lifted in a finite cover to a path whose initial and terminal points are in different components of the boundary.

DEFINITION 49. *A subgroup $H \le G$ is separable in G if for all $g \in G \setminus H$, G has a finite index normal subgroup N with $g \notin HN$.*

We can now state Hempel's criterion for the residual finiteness of a fundamental group of a graph of groups, which is [**Hem87**, Theorem 3.1].

THEOREM 50. *Let G be the fundamental group of a graph of groups with vertex set V and edge set E such that for each vertex group G_u and each $g \in G_u\backslash\{1\}$, there exist collections of finite index normal subgroups $\{N_v \le G_v\}_{v\in V}$ and $\{N_e \le G_e\}_{e\in E}$ which satisfy the following:*

(1) The collections $\{N_v\}_{v\in V}$ and $\{N_e\}_{e\in E}$ are compatible.
(2) $g \notin N_u$.
(3) If u is an endpoint of the edge e and $g \notin G_e$, then $g \notin G_eN_u$.

Then G is residually finite.

Hempel showed that Theorem 50 can be applied to any 3-dimensional extended graph manifold [**Hem87**], which, by the geometrization theorem, implies that all 3-manifolds have residually finite fundamental group (see Section 7 for more details).

Our aim is to apply this theorem to the fundamental groups of higher graph manifolds. A necessary condition for it to apply is that the boundary subgroups of each geometric piece are separable. We show this is indeed the case and, moreover, that boundary subgroups of any higher graph manifold M are separable in $\pi_1(M)$. Our main tool for this is the following lemma, which is a special case of a well-known algebraic result; see for example the final proposition in [**Lon87**].

LEMMA 51. *Let G be residually finite and let H be a maximal abelian or maximal nilpotent subgroup. Then H is separable in G.*

PROOF. Suppose that $g \in G\backslash H$. Define $[x_1, x_2] = x_1^{-1}x_2^{-1}x_1x_2$ and inductively define $[x_1, \dots, x_i] = [x_1, [x_2, \dots, x_i]]$ for all $x_1, \dots, x_i \in G$ and $i \ge 2$. That is, $[x_1, \dots, x_i]$ is the i-fold commutator. Since H is nilpotent, for some $k \ge 2$ and for all $x_1, \dots, x_k \in H$, $[x_1, \dots, x_k] = 1$. Moreover, since H is maximal nilpotent, there exists $g_1, \dots, g_k \in \langle H, g\rangle$ with $[g_1, \dots, g_k] \neq 1$. Since G is residually finite, there exists a finite index normal subgroup N with $[g_1, \dots, g_k] \notin N$. But the image of H in G/N is still nilpotent of the same degree, hence it cannot contain the image of g. Thus $g \notin HN$. $\square$

LEMMA 52. *Let M be an extended graph manifold or a pure complex-hyperbolic higher graph manifold with boundary. Then for each boundary component B, $\pi_1(B)$ is separable in $\pi_1(M)$.*

PROOF. We first consider the case where M is an extended graph manifold. In this case, B is the boundary component of some piece $N \cong V \times T^m$, where V is a compact manifold with toral boundary whose interior admits a finite volume real-hyperbolic metric, and $B = C \times T^m$, where C is a toral boundary component of V. Hence B is a torus and $\pi_1(B) \cong \pi_1(C) \times \pi_1(T^m)$, is abelian.

Let v be the vertex in the Bass–Serre tree stabilized by $\pi_1(N)$. If e is an edge adjacent to v, then its stabilizer is conjugate to $\pi_1(C') \times \pi_1(T^m)$ where $C' \times T^m$ is the boundary component corresponding to the edge e. $\pi_1(C)$ has trivial intersection with any conjugate of $\pi_1(C')$, since these are distinct maximal parabolic subgroups inside $\pi_1(V)$. Hence, if g is a nontrivial element of $\pi_1(C)$, then v is the unique vertex fixed by g. Now if h is any element of $\pi_1(M)$ which commutes with g, then $ghv = hgv = hv$, and since the fixed point of g is unique, we have $hv = v$. In particular, this implies that $h \in \pi_1(N)$. Now, since $\pi_1(C)$ is a maximal parabolic subgroup of $\pi_1(V)$, it is also a maximal abelian subgroup in $\pi_1(V)$, and hence we have that $\pi_1(C) \times \pi_1(T^m) \cong \pi_1(B)$ is a maximal abelian subgroup of $\pi_1(N)$. Hence $\pi_1(B)$ is also a maximal abelian in $\pi_1(M)$ and is thus separable in $\pi_1(M)$ by Lemma 51.

Now suppose that M is a pure complex-hyperbolic higher graph manifold. In this case, B is the boundary component of a piece W which is a compact manifold with nilmanifold boundary whose interior admits a finite volume complex-hyperbolic metric. As in the previous case v is the unique vertex fixed by any nontrivial element of $\pi_1(B)$. Let K be a nilpotent subgroup of $\pi_1(M)$ which contains $\pi_1(B)$. Let z be a nontrivial element of the center of K.

Since z commutes which each element of $\pi_1(B)$, z also fixes v, hence z is contained in $\pi_1(W)$. Since $\pi_1(B)$ is a maximal parabolic subgroup of $\pi_1(W)$, it is a maximal nilpotent subgroup of $\pi_1(W)$, so $z \in \pi_1(B)$. It follows that v is the unique fixed point of z and since each element of K commutes with z, we have that K stabilizes v and so $K \leq \pi_1(W)$. Since $\pi_1(B)$ is a maximal nilpotent subgroup of $\pi_1(W)$, $K = \pi_1(B)$ and so $\pi_1(B)$ is also maximal nilpotent in $\pi_1(M)$. Again, this implies that $\pi_1(B)$ is separable by Lemma 51. $\square$

While separability can be generalized to higher dimensions as above, Hempel's proof that compatible normal subgroups can be chosen is quite special to the 3-dimensional setting. The problem of funding such subgroups in higher dimensions is closely related to the cusp covering conjecture of Behrstock and Neumann [**BN12**]. We give a slightly stronger version of that conjecture since Behrstock and Neumann do not ask for the covers to be regular.

CONJECTURE 53 (Cusp covering conjecture). [**BN12**] *Let M be a hyperbolic n-manifold. Then for each cusp C of M there exists a sublattice Λ_C of $\pi_1(C)$ such that, for any choice of a sublattice $\Lambda'_C \subseteq \Lambda_C$ for each C, there exists a finite regular* *cover M' of M whose cusps covering each cusp C of M are the covers determined by Λ'_C.*

PROPOSITION 54. *Let M be a pure real-hyperbolic extended graph manifold. Then the cusp covering conjecture implies that $\pi_1(M)$ is residually finite.*

PROOF. Consider the splitting of $\pi_1(M)$ as a graph of groups. Let G_e be an edge group with adjacent vertex groups G_v and G_u. The cusp covering conjecture provides G_e with two finite index subgroups Λ_1 and Λ_2 where Λ_1 comes from considering G_e as a cusp subgroup of G_v and Λ_2 comes from considering G_e as a cusp subgroup of G_u. Define $\Lambda_e = \Lambda_1 \cap \Lambda_2$.

Now let G_u be a vertex group and let $g \in G_u \setminus \{1\}$. Since G_u is residually finite, there exists a finite index normal subgroup $K \leq G_u$ with $g \notin K$. For each adjacent edge group G_e embedded in G_u, there is a finite index normal subgroup K_e with $g \notin G_e K_e$ whenever $g \notin G_e$ (if $g \in G_e$ then we define $K_e = G_u$). Define K_u to be the intersection of K and each K_e. Then K_u is a finite index normal subgroup of G.

Now for each adjacent edge group G_e, the cusp covering conjecture says that G_u contains a finite index normal subgroup H with $H \cap G_e = \Lambda_e$. Finally, let $N_u = H \cap K_u$. Now N_u is a finite index normal subgroup of G such that $g \notin N_u$, $g \notin G_e N_u$ whenever $g \notin G_e$, and $N_u \cap G_e \leq \Lambda_e$ for all adjacent edges e. For each of these edges, let $N_e = N_u \cap G_e$.

Now for each edge group G_e which is not adjacent to G_v, we define $N_e = \Lambda_e$. For each other vertex group G_v, by the cusp covering conjecture there exists a finite index normal subgroup $N_v \leq G_v$ such that for each edge group G_e adjacent to G_v, $N_v \cap G_e = N_e$. Hence the collections $\{G_v\}_{v \in V}$ and $\{G_e\}_{e \in E}$ satisfy the hypothesis of Theorem 50, so $\pi_1(M)$ is residually finite. □

While Behrstock and Neumann only considered real-hyperbolic manifolds, there is a natural analogue of this conjecture for complex-hyperbolic manifolds. The proof of Proposition 54 can be adapted in the obvious way to show that this would imply that pure complex-hyperbolic higher graph manifolds have residually finite fundamental group.

PROPOSITION 55. *Let M be a pure complex-hyperbolic higher graph manifold. Then the complex-hyperbolic analogue of the cusp covering conjecture implies that $\pi_1(M)$ is residually finite.*

6.3. Iterated partial doubles. Without the cusp covering conjecture we are still able to construct infinite families of higher graph manifolds with residually finite fundamental group. This construction is based on the following algebraic lemma.

Let G be a group and $\mathcal{H} = \{H_1, \ldots, H_m\}$ a collection of subgroups of G. Let $G *_{\mathcal{H}} G$ denote the fundamental group of the graph of groups with two vertices both labeled by G and one edge labeled by each H_i.

LEMMA 56. *Let G be a residually finite group and $\mathcal{H}$ a finite collection of separable subgroups of G. Then $G *_{\mathcal{H}} G$ is residually finite.*

PROOF. We will show that G is residually finite using Theorem 50. To that end, let g be a nontrivial element of G. Since G is residually finite, there exists a finite index normal subgroup N_0 with $g \notin N_0$. Let $\mathcal{H} = \{H_1, \ldots, H_m\}$; by separability for each $1 \leq i \leq m$ there exists a finite index normal subgroup N_i with $g \notin H_i N_i$ whenever $g \notin H_i$ (if $g \in H_i$ then we define $N_i = G$). Define

$$N = \bigcap_{i=0}^{m} N_i$$

and note that N is a finite index normal subgroup of G. We now choose a copy of N inside each vertex group of $G *_{\mathcal{H}} G$ and choose $N \cap H_i$ in the edge group labeled by H_i. This is a compatible collection of subgroups for this splitting. Since $g \notin N_0$ and $N \subseteq N_0$, we have $g \notin N$. Similarly, whenever $g \notin H_i$ we have $g \notin H_i N_i$ and hence $g \notin H_i N$. Thus, N satisfies the hypothesis of Theorem 50, so $G *_{\mathcal{H}} G$ is residually finite. □

Let M be a higher graph manifold and let $\mathcal{B} = \{B_1, \ldots, B_m\}$ be a set of components of the boundary ∂M. We call the manifold obtained by taking two copies of M and gluing each B_i in one copy to the corresponding B_i in the other copy by the identity map a *partial double* of M. We denote this partial double by $M \cup_{\mathcal{B}} M$. If we let $G = \pi_1(M)$, and $H_i = \pi_1(B_i)$ and $\mathcal{H} = \{H_1, \ldots, H_m\}$, then $\pi_1(M \cup_{\mathcal{B}} M) \cong G *_{\mathcal{H}} G$. So, combining Lemma 52 and Lemma 56 we obtain the following theorem.

THEOREM 57. *Let M be an extended graph manifold or a pure complex-hyperbolic higher graph manifold with boundary and suppose $\pi_1(M)$ is residually finite. Let N be a partial double of M. Then $\pi_1(N)$ is residually finite.*

Theorem 57 can be applied inductively to produce infinite families of higher graph manifolds with residually finite fundamental groups. Explicitly, let V be a fixed finite volume real-hyperbolic manifold with at least two cusps. Let $\bar{V}$ be the manifold with boundary obtained by truncating the cusps of V. Let $\mathcal{B}_0$ be a *proper* subset of the components of $\partial \bar{V}$ and define $M_1 = \bar{V} \cup_{\mathcal{B}_0} \bar{V}$. M_1 is now an extended graph manifold with boundary with $\pi_1(M_1)$ residually finite by Theorem 57. Let $\mathcal{B}_1$ be a proper subset of the components of ∂M_1 and let $M_2 = M_1 \cup_{\mathcal{B}_1} M_1$ which, again, is an extended graph manifold with at least two boundary components and with $\pi_1(M_2)$ residually finite. Continuing inductively, we produce a sequence $M_1, M_2, \ldots$ of extended graph manifolds with at least two boundary components and residually finite fundamental group. We call these manifolds *iterated partial doubles* of V. For each M_i, let M_i' denote the double of M_i along the entire boundary of M_i. Then M_i' will be a closed extended graph manifold with residually finite fundamental group. Moreover, M_i' has 2^{i+1} geometric pieces which are all copies of V, so $\chi(M_i') = 2^{i+1}\chi(V)$. In particular, the manifolds $M_1', M_2', \ldots$ are all distinct. Thus, for each finite volume real-hyperbolic manifold with at least two cusps, we have an infinite family M_1', M_2' of distinct extended graph manifolds which satisfy the hypothesis of Theorem 3.

Starting with a finite volume complex-hyperbolic manifold W with at least two cusps, we can construct a sequence of iterated partial doubles M_1, M_2,...of W in the same way. Doubling each M_i along the entire boundary to produce closed manifolds M_i' as before, we obtain an infinite family M_1', M_2' of distinct pure complex-hyperbolic graph manifolds which satisfy the hypothesis of Theorem 4.

Given an extended graph manifold or a pure complex-hyperbolic graph manifold with boundary M, another way to produce a closed manifold is to start gluing components of the boundary of M together. Algebraically, this produces a manifold whose fundamental group is an HNN extension of $\pi_1(M)$. Preserving residual finiteness under such constructions is quite tricky but can be done in some special cases; see [**Hem76**, Chapter 15] where such constructions are studied in the context of 3-manifolds and [**BT78**] where they are studied from an algebraic perspective. One case where this construction will produce a manifold with residually finite fundamental group is the following.

Let G be a group with two collections of subgroups $\mathcal{H} = \{H_1, \ldots, H_m\}$ and $\mathcal{K} = \{K_1, \ldots, K_m\}$. Suppose α is a automorphism $\alpha\colon G \to G$ such that $\alpha(H_i) = K_i$ for each i. Consider the graph of groups with a single vertex with vertex group G and with edges $e_1, \ldots, e_m$ where the edge group for e_i is H_i attached via the identity map on one side and via the map $\alpha|_{H_i} : H_i \to K_i$ on the other. We denote the fundamental group of this graph of groups by $G*_{\mathcal{H},\alpha}$.

The following is an extended version of in [**BT78**, Lemma 4.4].

LEMMA 58. *Let G be a residually finite group with two collections of separable subgroups $\mathcal{H} = \{H_1, \ldots, H_m\}$ and $\mathcal{K} = \{K_1, \ldots, K_m\}$. Suppose α is a automorphism $\alpha\colon G \to G$ such that $\alpha(H_i) = K_i$ for each i. Then $G*_{\mathcal{H},\alpha}$ is residually finite.*

PROOF. Let $g \in G \setminus \{1\}$. Since G is residually finite, there exists a finite index normal subgroup N_0 with $g \notin N_0$. Since each H_i is separable, for each i we can find a finite index normal subgroup N_i such that if $g \notin H_i$ then $g \notin H_i N_i$ (whenever $g \in H_i$ we take $N_i = G$). Similarly choose finite index normal subgroups L_i such that if $g \notin K_i$ then $g \notin K_i L_i$. Now let $N' = N_0 \cap N_1 \cap \cdots \cap N_m \cap L_1 \cap \ldots L_m$, which is a finite index normal subgroup of G. Let N be the intersection of all subgroups of G with index $[G : N']$, and note that N is finite index, normal, and $\alpha(N) = N$ since α maps any subgroup to one of the same index.

Let $N_{e_i} = N \cap H_i$, and note that for each i the image of N_{e_i} in K_i is $\alpha(N_{e_i})$. We claim that $\{N\}$ and $\{N_{e_1}, \ldots, N_{e_m}\}$ are compatible collections of subgroups for this splitting of G. Clearly $N \cap H = N_e$, and also $N \cap K = \alpha(N) \cap \alpha(H) = \alpha(N \cap H) = \alpha(N_e)$. Since N is contained in N_0, each N_i, and each L_i, the collections $\{N\}$ and $\{N_2\}$ will also satisfy conditions (2) and (3) of Theorem 50. Hence the $G*_{\mathcal{H},\alpha}$ is residually finite by Theorem 50. □

An application of Lemma 58 together with Lemma 52 gives the following.

PROPOSITION 59. *Let M be an extended graph manifold or a pure complex-hyperbolic higher graph manifold with boundary and suppose $\pi_1(M)$ is residually finite. Let $\mathcal{B} = \{B_1, \ldots, B_m\}$ and $\mathcal{B}' = \{B_1', \ldots, B_m'\}$ be two collections of boundary components of M and suppose $f\colon M \to M$ is a homeomorphism with $f(B_i) = B_i'$ for each i. Then the manifold obtained by gluing each B_i to B_i' via the restriction of f has residually finite fundamental group.*

Let M be a manifold and let $\mathcal{B} = \{B_1, \ldots, B_m\}$ and $\mathcal{B}' = \{B_1', \ldots, B_m'\}$ be two collections of boundary components of M. Let $f\colon M \to M$ be a homeomorphism which maps each B_i to B_i'. Let $M \cup_{\mathcal{B},f} M$ denote the manifold obtained by taking two copies of M and gluing each B_i in the first copy to B_i' in the second via the restriction of f. We call $M \cup_{\mathcal{B},f} M$ a *twisted partial double* of M. Note that $M \cup_{\mathcal{B},f} M$ is a double cover of the manifold obtained from M by gluing each B_i to B_i'. Since finite index subgroups of residually finite groups are residually finite, we get the following corollary to Proposition 59.

Corollary 60. *Let M be an extended graph manifold or a pure complex-hyperbolic higher graph manifold with boundary and suppose $\pi_1(M)$ is residually finite. Then any twisted partial double of M has residually finite fundamental group.*

6.4. Explicit examples. In this section, we provide some explicit examples of four-dimensional pure complex-hyperbolic higher graph manifolds that are also iterated partial doubles. The literature seems to contain many examples of extended graph manifolds but, on the other hand, it seems to be lacking *any* pure complex-hyperbolic examples. The building blocks for our constructions are the explicit examples of cusped complex-hyperbolic surfaces with minimal volume constructed in [**DS18**]. We first present two examples, one with Euler characteristic two and the other with Euler characteristic four. Similar examples with arbitrarily large Euler characteristic can be easily manufactured. Indeed, as shown in [**DS17b**] one can construct explicit examples of cusped complex hyperbolic surfaces saturating the whole admissible volume spectrum, that is, with Euler characteristic equal to any positive integer. By using these examples with arbitrary large Euler characteristic, one can then construct explicit pure complex-hyperbolic higher graph manifolds with zero signature, with Euler number equal to any positive even number, and with residually finite fundamental group. Finally, we present a third example where the doubling operation is twisted via a self-isomorphism.

Example 61. *The first example is obtained by doubling along all its boundary components the complex hyperbolic surface with two (truncated) cusps given in* [**DS18**, Example 2]. *We denote this complex-hyperbolic surface by S. S has two cusps with cross sections S^1-bundles over tori, respectively, of Euler number one and three. We double gluing via the identity map, and if we consider the two copies of S with opposite orientations we then get an oriented 4-manifold with signature zero and Euler characteristic two. Let M_1 denote this manifold; see Figure 1.*

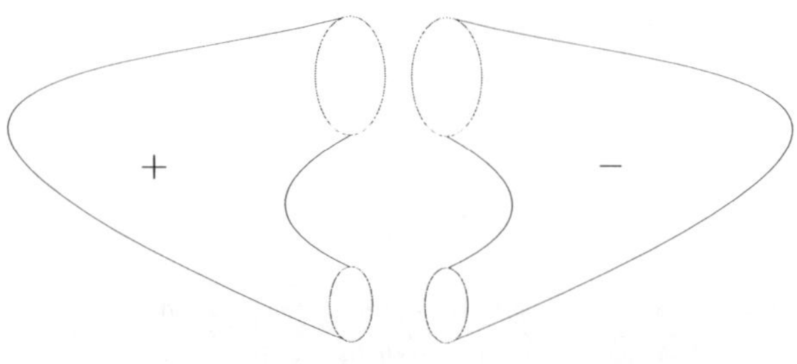

Figure 1. Two copies of S with opposite orientations. The red boundary components are S^1-bundles with Euler number 3 while the blue ones are S^1 bundles with Euler number 1. Adjacent boundary components are glued together via the identity map to produce M_1. The symbol "+" refers to the canonical orientation induced by the complex structure, and naturally the symbol "–" refers to the opposite orientation.

Example 62. *The second example is obtained by two iterated partial doubling operations on the complex hyperbolic surface with four cusps given in Example 1 of* [**DS18**] *(the so-called Hirzebruch's example* [**Hir84**]*). We denote this complex-hyperbolic surface by X. X has four cusps with cross sections S^1-bundles over tori* *all with Euler number one. We double first by gluing via the identity two boundary components, and then we double again to obtain a closed manifold. Again, if we consider the two vertices with the opposite orientation we then get an oriented 4-manifold with signature zero and Euler characteristic four which we denote by M_2; see Figure 2.*

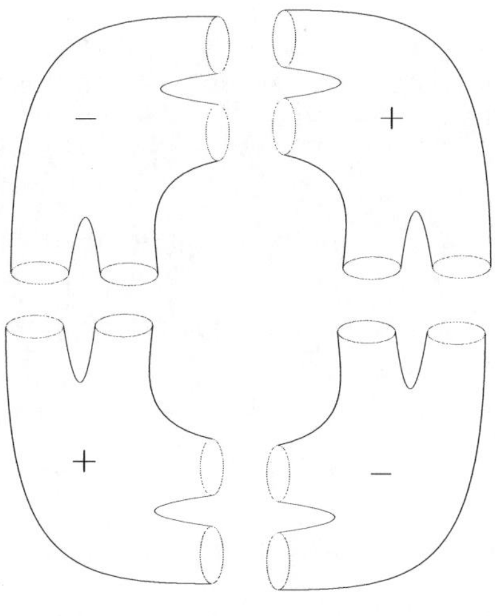

FIGURE 2. 4 copies of X with orientations as shown. Adjacent boundary components are glued together via the identity map to produce M_2.

EXAMPLE 63. *The third example is obtained again starting with X as in Example 62. In order to obtain a twisting self-isomorphism, we need to describe Hirzebruch's example in some more details. One starts with the elliptic curve E_ζ associated to the lattice $\mathbb{Z}[1,\zeta]$ where $\zeta = e^{\frac{\pi i}{3}}$. Inside the abelian surface $A = E_\zeta \times E_\zeta$ with coordinates (z,w), one considers the configuration of elliptic curves*

$$w = 0, \quad z = 0, \quad w = z, \quad w = \zeta z. \tag{64}$$

The abelian surface A admits a self-isomorphism $\psi : A \to A$ that preserves the configuration given in Equation (64) defined by setting

$$\psi(z,w) = (w, \zeta z).$$

More precisely, ψ maps the first elliptic curve in (64) to the second, the second to the first, the third to the fourth, and the fourth to the third. Now the configuration of elliptic curves in Equation (64) intersect in the point $(0,0) \in A$ only, and by blowing up this point we obtain a pair (X,D) where X is a blown-up abelian surface and where D is a divisor consisting of four smooth disjoint elliptic curves, say D_1, D_2, D_3, D_4. We then have that $\psi : A \to A$ lifts to a self-isomorphism $\bar{\psi} : (X,D) \to (X,D)$ where

$$\bar{\psi}(D_1) = D_2, \quad \bar{\psi}(D_2) = D_1, \quad \bar{\psi}(D_3) = D_4, \quad \bar{\psi}(D_4) = D_3, \quad \bar{\psi}(E) = E,$$

where E is the exceptional divisor in X. Now (X,D) is a smooth toroidal compactification of a complex-hyperbolic surface with Euler characteristic one and with *four cusps; see* [**DS18**] *for more details. The cusps are in one-to-one correspondence with the elliptic curves* D_i, $i = 1, 2, 3, 4$, *and the complex-hyperbolic surface is biholomorphic to* $X \setminus D$. *Thus,* $\bar{\psi} : X \setminus D \to X \setminus D$ *is a biholomorphism preserving the punctured exceptional divisor which is totally geodesic in* $(X \setminus D, g_{\mathbb{C}})$ *(see* [**DS19b**, Proposition 2.1]*).*

Now as in Example 61, $X \setminus D$ *has four cusps with cross sections* S^1*-bundles over tori all with Euler number one. We can then produce a twisted double by using* $\bar{\psi}$. *We denote this manifold by* M_3. *Again, if we consider the two copies of* X *with the opposite orientation we then get an oriented* 4*-manifold with signature zero and Euler characteristic two; see Figure* 3.

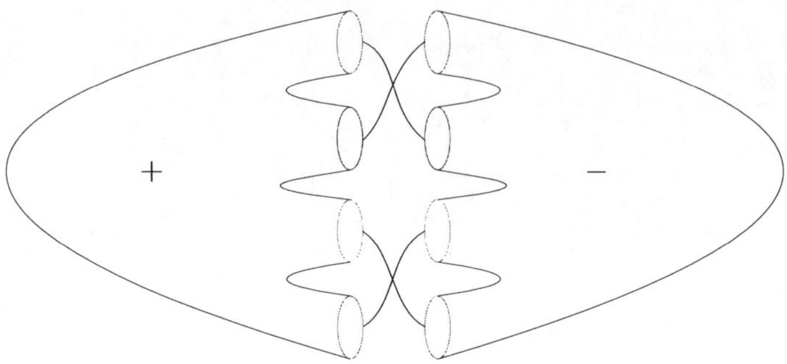

FIGURE 3. 2 copies of X with opposite orientations. Boundary components are glued together via the restriction of $\bar{\psi}$ according to the given matching to produce M_3.

REMARK 65. *Theorems* 4 *and* 57 *immediately imply that the closed* 4*-manifolds constructed in Examples* 61 *and* 62 *do satisfy the Singer conjecture. Recall that because of* [**CSS19**, Lemma 14], *both examples are indeed aspherical.*

7. The Singer conjecture in dimension three revisited

The Singer conjecture was proved for 3-manifolds in a celebrated paper of Lück and Lott. Here we show that Theorem 3 can be used to give another proof.

We first recall some 3-manifold topology; for more details, see [**AFW15**] and [**Thu80**]. A closed 3-manifold M is aspherical if and only if $\pi_1(M)$ is infinite and M is irreducible, that is every embedded 3-sphere bounds a 3-ball in M. Passing to a double cover of M if necessary, we can assume that M is orientable. To any such M we can apply the geometrization theorem, see [**AFW15**, Theorem 1.7.6] for the version we state here.

THEOREM 66 (Geometrization theorem (Perelman)). *Let* M *be a closed, irreducible, orientable 3-manifold. Then* M *has a collection of disjoint, embedded, incompressible tori* $\{T_1, \dots, T_m\}$ *such that each component of* $M \setminus \{T_1, \dots, T_m\}$ *is either Seifert-fibered or hyperbolic.*

We next use the geometrization theorem to show that, except in a few special cases, closed aspherical 3-manifolds are finitely covered by extended graph manifolds.

LEMMA 67. *Let M be a closed, aspherical 3-manifold. Suppose M is not Seifet-fibered and not finitely covered by a torus bundle over a circle. Then M is finitely covered by an extended graph manifold.*

PROOF. Passing to a double cover if necessary, we can assume M is orientable. By the geometrization theorem, M has a decomposition along tori into pieces which are either hyperbolic or Seifert-fibered. Some of these pieces may be twisted I-bundles over Klein bottles, but we can pass to a finite cover of M and refine the collection of tori we are cutting along such that no such pieces occur. In particular, after passing to a finite cover of M we can assume that each Seifert-fibered piece in the decomposition has a hyperbolic base orbifold (see, for example, [**GHL21**, Lemma 3.8]). Let $P_1, \ldots, P_k$ denote the hyperbolic pieces and $Q_1,\ldots,Q_l$ denote the Seifert-fibered pieces. For each Q_i, as in [**Hem87**, Lemma 4.1] we can find a finite cover $\overline{Q}_i \cong S^1 \times \Sigma_i$ of Q_i where Σ_i is a compact hyperbolic surface and each boundary component of $\overline{Q}_i$ projects homeomorphically to a boundary component of Q. Considering each P_i as a cover of itself, these will form a compatible collection of covers. Hence by Theorem 48 there is a finite cover $\overline{M}$ of M such that $\overline{M}$ is a graph of spaces with tori edge spaces and each vertex space is hyperbolic or is a copy of $S^1 \times \Sigma_i$ for some i. Hence $\overline{M}$ is an extended graph manifold. □

THEOREM 68 (Lott and Lück [**LL95**] & geometrization). *The Singer conjecture holds for closed aspherical 3-manifolds.*

PROOF. Let M be a closed, aspherical 3-manifold. If M is Seifert-fibered or has a geometry modeled on torus bundle over the circle with Anosov gluing (the so-called Sol-geometry), then $MinVol(M) = 0$ and $\pi_1(M)$ is residually finite. For a proof of these facts, we refer to [**CG86**, Examples 0.2 and 1.2]. Notice that Example 0.2 is continued in [**CG86**, page 311]. By Proposition 40, the L^2-Betti numbers of M vanish. Lemma 67 shows that in all other cases, M is finitely covered by an extended graph manifold. $\pi_1(M)$ is residually finite by [**Hem87**] and hence the L^2-Betti numbers vanish by Theorem 3. □

REMARK 69. *Alternatively in the proof of Theorem* 68, *if M is Seifert-fibered or has a geometric Sol-structure, then $\pi_1(M)$ has an infinite normal abelian subgroup and hence the L^2–Betti numbers vanish by a result of Cheeger and Gromov; see* [**CG86b**, Corollary 0.6].

Acknowledgments

The first author thanks M. Anderson, G. Besson, and S. Maillot for introducing him to the theory of geometric 3-manifolds during his time as a graduate student. He also thanks M. Stern for a stimulating ongoing collaboration on Price inequalities for harmonic forms, and for pointing out a mistake in a previous version of this paper. The second author thanks D. Groves, D. B. McReynolds, and E. Einstein for helpful conversations about separability and virtually special groups. The second author also thanks the University of Florida, which he visited during the preparation of this work, for its hospitality and support. Finally, the authors thank C. Löh for pointing out reference [**FFL19**], and the referees for pertinent comments on the paper.

References

[Ago13] I. Agol, *The virtual Haken conjecture*, Doc. Math. **18** (2013), 1045–1087. With an appendix by Agol, Daniel Groves, and Jason Manning. MR3104553

[AFW15] M. Aschenbrenner, S. Friedl, and H. Wilton, *3-manifold groups*, EMS Series of Lectures in Mathematics, European Mathematical Society (EMS), Zürich, 2015, DOI 10.4171/154. MR3444187

[Ati76] M. F. Atiyah, *Elliptic operators, discrete groups and von Neumann algebras*, Colloque "Analyse et Topologie" en l'Honneur de Henri Cartan (Orsay, 1974), Astérisque, No. 32-33, Soc. Math. France, Paris, 1976, pp. 43–72. MR420729

[BT78] B. Baumslag and M. Tretkoff, *Residually finite HNN extensions*, Comm. Algebra **6** (1978), no. 2, 179–194, DOI 10.1080/00927877808822240. MR484178

[BN12] J. A. Behrstock and W. D. Neumann, *Quasi-isometric classification of non-geometric 3-manifold groups*, J. Reine Angew. Math. **669** (2012), 101–120, DOI 10.1515/crelle.2011.143. MR2980453

[BW12] N. Bergeron and D. T. Wise, *A boundary criterion for cubulation*, Amer. J. Math. **134** (2012), no. 3, 843–859, DOI 10.1353/ajm.2012.0020. MR2931226

[BHW11] N. Bergeron, F. Haglund, and D. T. Wise, *Hyperplane sections in arithmetic hyperbolic manifolds*, J. Lond. Math. Soc. (2) **83** (2011), no. 2, 431–448, DOI 10.1112/jlms/jdq082. MR2776645

[Bor63] A. Borel, *Compact Clifford-Klein forms of symmetric spaces*, Topology **2** (1963), 111–122, DOI 10.1016/0040-9383(63)90026-0. MR146301

[CHK20] I. Chatterji, S. Hughes, and P. Kropholler, *The first ℓ^2-Betti number and groups acting on trees*, Proc. Edinb. Math. Soc. (2) **64** (2021), no. 4, 916–923, DOI 10.1017/S0013091521000663. MR4349416

[CG85] J. Cheeger and M. Gromov, *Bounds on the von Neumann dimension of L^2-cohomology and the Gauss-Bonnet theorem for open manifolds*, J. Differential Geom. **21** (1985), no. 1, 1–34. MR806699

[CG86] J. Cheeger and M. Gromov, *Collapsing Riemannian manifolds while keeping their curvature bounded. I*, J. Differential Geom. **23** (1986), no. 3, 309–346. MR852159

[CG86b] J. Cheeger and M. Gromov, *L_2-cohomology and group cohomology*, Topology **25** (1986), no. 2, 189–215, DOI 10.1016/0040-9383(86)90039-X. MR837621

[CSS19] C. Connell and P. Suárez-Serrato, *On higher graph manifolds*, Int. Math. Res. Not. IMRN **5** (2019), 1281–1311, DOI 10.1093/imrn/rnx133. MR3920347

[Dah03] F. Dahmani, *Combination of convergence groups*, Geom. Topol. **7** (2003), 933–963, DOI 10.2140/gt.2003.7.933. MR2026551

[DD17] G. Di Cerbo and L. F. Di Cerbo, *On the canonical divisor of smooth toroidal compactifications*, Math. Res. Lett. **24** (2017), no. 4, 1005–1022, DOI 10.4310/MRL.2017.v24.n4.a4. MR3723801

[DS17b] L. F. Di Cerbo and M. Stover, *Bielliptic ball quotient compactifications and lattices in* PU(2, 1) *with finitely generated commutator subgroup* (English, with English and French summaries), Ann. Inst. Fourier (Grenoble) **67** (2017), no. 1, 315–328, DOI 10.5802/aif.3083. MR3612333

[DS18] L. F. Di Cerbo and M. Stover, *Classification and arithmeticity of toroidal compactifications with $3\bar{c}_2 = \bar{c}_1^2 = 3$*, Geom. Topol. **22** (2018), no. 4, 2465–2510, DOI 10.2140/gt.2018.22.2465. MR3784527

[DL19] L. F. Di Cerbo and L. Lombardi, *Moving Seshadri constants, and coverings of varieties of maximal Albanese dimension*, Asian J. Math. **25** (2021), no. 2, 305–320, DOI 10.4310/AJM.2021.v25.n2.a8. MR4334419

[DS19] L. F. Di Cerbo and M. Stern, *Harmonic forms, Price inequalities, and Benjamini-Schramm convergence*, Preprint, `arXiv:1909.05634v1 [math.DG]`, (2019).

[DS19b] L. F. Di Cerbo and M. Stover, *Punctured spheres in complex hyperbolic surfaces and bielliptic ball quotient compactifications*, Trans. Amer. Math. Soc. **372** (2019), no. 7, 4627–4646, DOI 10.1090/tran/7650. MR4009437

[DS20] L. F. Di Cerbo and M. Stern, *On the Betti numbers of finite volume hyperbolic manifolds*, To appear in J. Differential Geom., `arXiv:2009.11431v3 [math.DG]`, (2020).

[DS22] L. F. Di Cerbo and M. Stern, *Price inequalities and Betti number growth on manifolds without conjugate points*, Comm. Anal. Geom. **30** (2022), no. 2, 297–334. MR4516072

[Ebe80] P. Eberlein, *Lattices in spaces of nonpositive curvature*, Ann. of Math. (2) **111** (1980), no. 3, 435–476, DOI 10.2307/1971104. MR577132

[FFL19] D. Fauser, S. Friedl, and C. Löh, *Integral approximation of simplicial volume of graph manifolds*, Bull. Lond. Math. Soc. **51** (2019), no. 4, 715–731, DOI 10.1112/blms.12266. MR3990387

[FV17] T. Fernós and A. Valette, *The Mayer-Vietoris sequence for graphs of groups, property (T), and the first ℓ^2-Betti number*, Homology Homotopy Appl. **19** (2017), no. 2, 251–274, DOI 10.4310/HHA.2017.v19.n2.a13. MR3723770

[F21] S. Fisher, *Improved algebraic fibrings*, Preprint, `arXiv:2112.00397`, (2021).

[FLS15] R. Frigerio, J.-F. Lafont, and A. Sisto, *Rigidity of high dimensional graph manifolds* (English, with English and French summaries), Astérisque **372** (2015), xxi+177. MR3444648

[Fuk89] K. Fukaya, *Collapsing Riemannian manifolds to ones with lower dimension. II*, J. Math. Soc. Japan **41** (1989), no. 2, 333–356, DOI 10.2969/jmsj/04120333. MR984756

[Gol99] W. M. Goldman, *Complex hyperbolic geometry*, Oxford Mathematical Monographs, The Clarendon Press, Oxford University Press, New York, 1999. Oxford Science Publications. MR1695450

[Gro82] M. Gromov, *Volume and bounded cohomology*, Inst. Hautes Études Sci. Publ. Math. **56** (1982), 5–99 (1983). MR686042

[Gro99] M. Gromov, *Metric structures for Riemannian and non-Riemannian spaces*, Progress in Mathematics, vol. 152, Birkhäuser Boston, Inc., Boston, MA, 1999. Based on the 1981 French original [MR0682063 (85e:53051)]; With appendices by M. Katz, P. Pansu and S. Semmes; Translated from the French by Sean Michael Bates. MR1699320

[GHL21] D. Groves, M. Hull, and H. Liang, *Homomorphisms to 3-manifold groups*, Preprint, `arXiv:2103.06095v1 [math.GT]`, (2021).

[Har71] G. Harder, *A Gauss-Bonnet formula for discrete arithmetically defined groups*, Ann. Sci. École Norm. Sup. (4) **4** (1971), 409–455. MR309145

[Hem76] J. Hempel, 3-*Manifolds*, Annals of Mathematics Studies, No. 86, Princeton University Press, Princeton, NJ; University of Tokyo Press, Tokyo, 1976. MR415619

[Hem87] J. Hempel, *Residual finiteness for* 3-*manifolds*, Combinatorial group theory and topology (Alta, Utah, 1984), Ann. of Math. Stud., vol. 111, Princeton Univ. Press, Princeton, NJ, 1987, pp. 379–396. MR895623

[Hir84] F. Hirzebruch, *Chern numbers of algebraic surfaces: an example*, Math. Ann. **266** (1984), no. 3, 351–356, DOI 10.1007/BF01475584. MR730175

[HW19] J. Huang and D. Wise, *Stature and separability in graphs of groups*, Preprint, `arXiv:1904.06021`, (2019).

[Hru10] G. C. Hruska, *Relative hyperbolicity and relative quasiconvexity for countable groups*, Algebr. Geom. Topol. **10** (2010), no. 3, 1807–1856, DOI 10.2140/agt.2010.10.1807. MR2684983

[HW14] G. C. Hruska and D. T. Wise, *Finiteness properties of cubulated groups*, Compos. Math. **150** (2014), no. 3, 453–506, DOI 10.1112/S0010437X13007112. MR3187627

[LS00] B. Leeb and P. Scott, *A geometric characteristic splitting in all dimensions*, Comment. Math. Helv. **75** (2000), no. 2, 201–215, DOI 10.1007/PL00000370. MR1774701

[Lon87] D. D. Long, *Immersions and embeddings of totally geodesic surfaces*, Bull. London Math. Soc. **19** (1987), no. 5, 481–484, DOI 10.1112/blms/19.5.481. MR898729

[LK08] B. Kleiner and J. Lott, *Notes on Perelman's papers*, Geom. Topol. **12** (2008), no. 5, 2587–2855, DOI 10.2140/gt.2008.12.2587. MR2460872

[Lüc94] W. Lück, *Approximating L^2-invariants by their finite-dimensional analogues*, Geom. Funct. Anal. **4** (1994), no. 4, 455–481, DOI 10.1007/BF01896404. MR1280122

[Lüc02] W. Lück, *L^2-invariants: theory and applications to geometry and K-theory*, Ergebnisse der Mathematik und ihrer Grenzgebiete. 3. Folge. A Series of Modern Surveys in Mathematics [Results in Mathematics and Related Areas. 3rd Series. A Series of Modern Surveys in Mathematics], vol. 44, Springer-Verlag, Berlin, 2002, DOI 10.1007/978-3-662-04687-6. MR1926649

[LL95] J. Lott and W. Lück, *L^2-topological invariants of* 3-*manifolds*, Invent. Math. **120** (1995), no. 1, 15–60, DOI 10.1007/BF01241121. MR1323981

[Pie19] E. Pieroni, *Minimal entropy of 3-manifolds*, PhD Thesis, University of Rome "La Sapienza", `arXiv:1902.09190v1 [math.DG]`, (2018).

[Ron93] X. Rong, *The limiting eta invariants of collapsed three-manifolds*, J. Differential Geom. **37** (1993), no. 3, 535–568. MR1217160

[Sau09] R. Sauer, *Amenable covers, volume and L^2-Betti numbers of aspherical manifolds*, J. Reine Angew. Math. **636** (2009), 47–92, DOI 10.1515/CRELLE.2009.082. MR2572246

[Sou01] J. Souto, *Minimal volume and minimal entropy*, PhD Thesis, Part II. Bonn University, 2001.

[Thu80] W. P. Thurston, *Geometry and topology of three-manifolds*, Princeton lecture notes available at `http://www.msri.org/publications/books/gt3m/`, 1980.

[Wis21] D. T. Wise, *The structure of groups with a quasiconvex hierarchy*, Annals of Mathematics Studies, vol. 209, Princeton University Press, Princeton, NJ, [2021] ©2021. MR4298722

[Wol11] J. A. Wolf, *Spaces of constant curvature*, 6th ed., AMS Chelsea Publishing, Providence, RI, 2011, DOI 10.1090/chel/372. MR2742530

Department of Mathematics, University of Florida, Gainesville, Floria 32607
Email address: `dicerbo@ufl.edu`

Department of Mathematics & Statistics, University of North Carolina, Greensboro, North Carolina 27402
Email address: `mbhull@uncg.edu`

Contemporary Mathematics
Volume **816**, 2025
https://doi.org/10.1090/conm/816/16356

L_p-cohomology and the geometry of p-harmonic forms

Mark Stern

ABSTRACT. In this note we describe basic geometric properties of p-harmonic forms and p-coclosed forms and use them to reprove vanishing theorems of Pansu and new injectivity theorems for the L_p-cohomology of simply connected, pinched negatively curved manifolds. We also provide a partial resolution of a conjecture of Gromov on the vanishing of L_p-cohomology on symmetric spaces.

1. Introduction

Let (M^n, g) be a complete Riemannian n-manifold. Let

$$L_p^k(M) := \{\text{measurable } k\text{-forms } f : f \in L_p\},$$

$$C_p^k(M) := \{f : f \in L_p^k(M) \text{ and } df \in L_p^{k+1}(M)\},$$

$$Z_p^k(M) := \{z \in C_p^k(M) : dz = 0\},$$

and

$$B_p^k(M) := dC_p^{k-1}(M).$$

Define the L_p-cohomology to be the quotient

$$H_p^k(M) := Z_p^k(M)/B_p^k(M).$$

When M is compact, the de Rham cohomology and the L_p-cohomology are isomorphic, for all $1 < p < \infty$. For noncompact manifolds, however, these cohomology groups need not coincide. Moreover, in the noncompact case, for some p and k, $B_p^k(M)$ may not be a closed subspace of $Z_p^k(M)$. In this case, the L_p-cohomology naturally decomposes into two (possibly trivial) summands: the reduced cohomology

$$H_{p,red}^k(M) := Z_p^k(M)/\bar{B}_p^k(M),$$

and the (either zero or infinite dimensional non-Hausdorff) torsion cohomology

$$T_p^k(M) := \bar{B}_p^k(M)/B_p^k(M),$$

where $\bar{B}_p^k(M)$ denotes the L_p closure of $B_p^k(M)$. Given $1 < p < \infty$, and $0 \leq k \leq n$, we define the space of p-harmonic k-forms to be

$$\mathcal{H}_p^k(M) := \{h \in Z_p^k(M) : d^*(|h|^{p-2}h) = 0\}.$$

2020 *Mathematics Subject Classification.* Primary 58A14; Secondary 57T15.
The author was supported in part by Simons Foundation grant 391000626.

The p-harmonics play the role in L_p-cohomology that the L^2-harmonics play in L^2-cohomology: coupling the geometry and topology and providing simple tools for proving vanishing theorems.

The purpose of this note is to develop the elementary properties of the p-harmonics and apply them to the study of L_p-cohomology. In Section 2, we collect in one place some basic elements of L_p-Hodge theory. Most of these somewhat elementary results are well known to analysts but perhaps less familiar to topologists. We prove the existence of canonical L_p-primitives, the isomorphism between $H^k_{p,red}(M)$ and $\mathcal{H}^k_p(M)$, and the $\mathcal{H}^k_p(M) \leftrightarrow \mathcal{H}^{n-k}_{\frac{p}{p-1}}(M)$ Poincaré duality defined by a nonlinear L_p-Hodge star operator. In Section 3, we derive a Bochner formula for p-harmonics. This formula is well known, but rarely labeled or exploited as a Bochner formula in the literature. The usual package of vanishing theorems one obtains from the Bochner formula in the L^2 case extends immediately to all $p \in (1, \infty)$.

In Section 4, we examine the extension of the monotonicity results of [**DS22**] to p-harmonics. Given a p-harmonic form, h_p, and $x \in M$, define for $R < \mathrm{inj}_M$ (the injectivity radius of M),

$$I_p(h_p, x, R) := \int_{B_R(x)} |h_p|^p dv,$$

where $B_R(x)$ denotes the geodesic ball of radius R and center x. In [**DS22**], we showed that for M compact, $\delta < R \leq \mathrm{inj}_M$, and for some constant $c(\tilde{M}) > 0$, depending only on the universal cover $\tilde{M}$ of M, we have

$$\frac{b^k(M)}{vol(M)} \leq \sup_{h_2 \in \mathcal{H}^k_2(M)\setminus\{0\}} c(\tilde{M}) \sup_{x \in M} \frac{\delta^{-n} I_2(h_2, x, \delta)}{I_2(h_2, x, R)}. \tag{1.1}$$

Here $b^k(M)$ denotes the kth betti number of M. If the sectional curvature of M, $\sec_M$, satisfies

$$-1 \leq \sec_M < -\frac{k^2}{(n-k-1)^2},$$

then we also showed in [**DS22**] that

$$\frac{I_2(h_2, x, \delta)}{I_2(h_2, x, R)}$$

is exponentially small in $R - \delta$. This yields 2 new proofs of the vanishing of $\mathcal{H}^k_2(\tilde{M})$, when M has residually finite fundamental group and

$$-1 \leq \sec_M < -\frac{k^2}{(n-k-1)^2}.$$

One proof follows from applying (1.1) to a tower of covers and applying the Lück approximation theorem ([**Lu94**]). The other proof (which does not require residual finiteness) follows immediately from the vanishing of

$$\frac{I_2(h_2, x, \delta)}{\|h_2\|^2_{L^2(\tilde{M})}} = \frac{I_2(h_2, x, \delta)}{I_2(h_2, x, \infty)} = 0.$$

Recall that the Singer conjecture for compact negatively curved manifolds M posits that

$$\mathcal{H}^k_2(\tilde{M}) = 0, \text{ for } k \neq \frac{dim(M)}{2}.$$

Hence we are very interested in relaxing the pinching conditions required for applying [**DS22**] to prove $\mathcal{H}_2^k(\tilde{M}) = 0$.

In Section 5, we show that if

$$-1 \leq \sec_M \leq -\frac{(p-1)k}{n-k-1}, \quad \text{or} \quad -1 \leq \sec_M \leq -\frac{n-k}{(k-1)(p-1)},$$

then $\frac{I_p(h_p,x,\delta)}{I_p(h_p,x,R)}$ is again exponentially small in $R-\delta$. Observe that as p tends to 1 or ∞, the pinching condition limits to the assumption of strictly negative curvature. (See Theorem 5.6.) Unfortunately, this does not immediately imply the Singer conjecture, because for $p \neq 2$, the p-harmonic forms induce a Banach space but not a Hilbert space structure on the L_p-cohomology, and the Hilbert space structure is used to deduce (1.1). On the other hand, exponential decay of the ratio $\frac{I_p(h_p,x,\delta)}{I_p(h_p,x,R)}$ implies $\mathcal{H}_p^k(\tilde{M}) = 0$. This gives a "harmonic" proof of Pansu's vanishing theorems for reduced L_p-cohomology of complete simply connected manifolds for

$$p < \delta\frac{n-k-1}{k} + 1 \quad \text{and dually for} \quad p > \frac{n-k}{\delta(k-1)} + 1.$$

See [**Pan08**]. In Section 6, we extend these estimates to p-coclosed forms (as defined in Definition 2.8) in order to reproduce Pansu's vanishing theorems [**Pan08**] for the torsion summand of the L_p-cohomology, for $p < \delta\frac{n-k}{k-1} + 1$.

In Section 7, we consider the (necessarily reduced) L_p-cohomology for $p < \delta\frac{n-k}{k-1} + 1$, and prove injectivity results for the natural mapping from $H_p^k(M) \to H_q^k(M)$ for $p \leq q$ and $q - p$ small.

In Section 8, we turn from manifolds of pinched negative curvature to symmetric spaces, and consider the following conjecture of Gromov [**Gro85**, Section 8C1].

CONJECTURE 1.2. *If G is a semisimple Lie group, with maximal compact subgroup K, then $H_p^k(G/K) = 0$, for $1 < p < \infty$, and $k < \operatorname{rank}_{\mathbb{R}}(G)$.*

As an elementary application of the techniques for analyzing p-harmonic forms and p-coclosed forms developed in the proceeding sections, we prove the following theorem, which provides a partial resolution of this conjecture.

THEOREM 1.3. *If G is a simple Lie group, with maximal compact subgroup K, then $H_p^k(G/K) = 0$, for $1 < p \leq 2$, and $k < \operatorname{rank}_{\mathbb{R}}(G)$.*

2. The nonlinear Hodge theorem and p-harmonic forms

One form of an L_p-Hodge theorem is a decomposition of L_p forms into

$$\overline{Im(d)} \oplus Ker(\Delta) \oplus \overline{Im(d^*)}, \tag{2.1}$$

with domains defined appropriately on L_p forms. See for example [**Sco95**]. Such results are fundamental, and if verified in our context would imply a significant extension of our theorems. They do not, however, exploit many of the geometric properties that make L_p-cohomology interesting and distinct from L^2-cohomology.

For our purposes, p-harmonic forms will replace $Ker(\Delta)$ in the above decomposition and p-coclosed primitives (defined below in Definition 2.8) will replace primitives chosen from $\overline{Im(d^*)}$. Much of the content of this section is already apparent in [**SS70**]. See also [**KLV21**, 11.2] for a clear and careful treatment of related theorems for functions and [**ISS99**] for both linear and nonlinear approaches to Hodge theory.

As a first step toward the Hodge theorem, we show the existence of canonical primitives for elements of $B_p^k(M)$. In the following, M is a complete manifold, possibly with boundary.

LEMMA 2.2. *For each $\zeta \in B_p^k(M)$, $\exists! \beta_\infty \in C_p^{k-1}(M)$ such that $d\beta_\infty = \zeta$, and*

$$d^* \left(|\beta_\infty|^{p-2}\beta_\infty\right) = 0.$$

PROOF. Write $d^{-1}(\zeta)$ for the set $\{\beta \in C_p^{k-1}(M) : d\beta = \zeta\}$. Set $\nu(\zeta) := \inf_{\beta\in d^{-1}(\zeta)} \|\beta\|_{L_p}^p$. Let $\{\beta_j\}_{j=1}^\infty \subset d^{-1}(\zeta)$ satisfy $\lim_{j\to\infty} \|\beta_j\|_{L_p}^p = \nu(\zeta)$. Passing to a subsequence if necessary, we can assume $\{\beta_j\}_{j=1}^\infty$ converges weakly in L_p to some $\beta_\infty \in L_p$. Since d commutes with weak limits, we have $d\beta_\infty = \zeta$. Let ϕ be a smooth compactly supported $(k-2)$-form and $t \in \mathbb{R}$. Then

$$\begin{aligned} 0 &\leq \int_M (|\beta_\infty + td\phi|^p - |\beta_\infty|^p)dv \\ &= \int_M \int_0^1 \frac{d}{ds}|\beta_\infty + std\phi|^p ds dv \\ &= p\int_M \int_0^1 |\beta_\infty + std\phi|^{p-2}\langle\beta_\infty + std\phi, td\phi\rangle ds dv. \end{aligned} \tag{2.3}$$

Choose $t = -\frac{1}{n}\int_M |\beta_\infty|^{p-2}\langle\beta_\infty, d\phi\rangle dv =: \frac{c}{n}$, $n \in \mathbb{N}$. Then we have

$$0 \leq -p\int_M |\beta_\infty|^{p-2}\langle\beta_\infty, d\phi\rangle dv \int_M \int_0^1 |\beta_\infty + s\frac{c}{n}d\phi|^{p-2}\langle\beta_\infty + s\frac{c}{n}d\phi, d\phi\rangle ds dv. \tag{2.4}$$

Taking the limit as $n \to \infty$ yields

$$0 \leq -p(\int_M |\beta_\infty|^{p-2}\langle\beta_\infty, d\phi\rangle dv)^2. \tag{2.5}$$

Hence for all ϕ smooth and compactly supported,

$$0 = \int_M |\beta_\infty|^{p-2}\langle\beta_\infty, d\phi\rangle dv, \tag{2.6}$$

and therefore

$$d^* \left(|\beta_\infty|^{p-2}\beta_\infty\right) = 0. \tag{2.7}$$

Since the L_p norm is strictly convex, it has a unique critical point. Hence β_∞ is unique. □

DEFINITION 2.8. We call the primitive specified in Lemma 2.2 the *p-coclosed primitive*.

For $z \in Z_p^*(M)$, let $[z]$ and $[z]_{red}$ denote the images of z in $H_p^*(M)$ and $H_{p,red}^*(M)$, respectively.

THEOREM 2.9 (Nonlinear Hodge theorem). *For each $\varphi \in H_{p,red}^k(M)$, $\exists! h \in Z_p^k(M)$ such that $[h]_{red} = \varphi$, and*

$$d^*(|h|^{p-2}h) = 0. \tag{2.10}$$

Moreover, $\|h\|_{L_p(M)} \leq \|z\|_{L_p(M)}$, for all $z \in Z_p^k(M)$ such that $[z]_{red} = \varphi$, with equality only if $h = z$.

PROOF. Let $z \in Z_p^k(M)$. Define $E_z : \bar{B}_p^k(M) \to \mathbb{R}$ by

$$E_z(\beta) := \int_M |z - \beta|^p dv.$$

Let $\rho(z) := \inf_{\beta \in \bar{B}_p^k(M)} E_z(\beta)$. Then $\rho(z) = 0$ if and only if $[z]_{red} = 0$. Suppose $\rho(z) \neq 0$. Let $\{\beta_j\}_{j=1}^{\infty} \subset \bar{B}_p^k(M)$ satisfy $\lim_{j\to\infty} E_z(\beta_j) = \rho(z)$. Then $\{\beta_j\}_{j=1}^{\infty}$ is an L_p bounded sequence. Passing to a subsequence, we can assume $\{\beta_j\}_{j=1}^{\infty}$ is weakly convergent to some $\zeta \in \bar{B}_p^k(M)$. By the weak lower semicontinuity of convex functions, $E_z(\zeta) \leq \rho(z)$, and therefore $E_z(\zeta) = \rho(z)$. Set $h := z - \zeta$. We then have

$$[h]_{red} = [z]_{red}.$$

The proof that $d^*(|h|^{p-2}h) = 0$ and that h is the unique norm minimizer in its class is now the same as the proofs of the corresponding statements in Lemma 2.2. □

COROLLARY 2.11. *Let M be a complete riemannian manifold. Then*

$$H_{p,red}^k(M) \cong \mathcal{H}_p^k(M).$$

We can also dualize our treatment of L_p Hodge theory. Define

$$\Omega_p^k(M) := \{\text{measurable } k-\text{forms } f : f \in L_p \text{ and } d^*f \in L_p\}.$$

$$F_p^k(M) := \{f \in \Omega_p^k(M) : d^*f = 0\}, \text{ and } D_p^k(M) := d^*\Omega_p^{k+1}(M).$$

Define the L_p-homology to be

$$Y_p^k(M) := F_p^k(M)/D_p^k(M).$$

Define the torsion summand

$$\mathcal{T}_p^k(M) := \bar{D}_p^k(M)/D_p^k(M),$$

and the reduced homology

$$Y_{p,\mathrm{red}}^k(M) := F_p^k(M)/\bar{D}_p^k(M).$$

Define the space of p-coharmonic forms to be

$$\mathcal{E}_p^k(M) := \{f \in F_p^k(M) : d(|f|^{p-2}f) = 0\}.$$

For complete manifolds, we again have that $Y_{p,\mathrm{red}}^k(M)$ is isomorphic to $\mathcal{E}_p^k(M)$.

Analytic Poincaré duality proofs survive the passage from 2−harmonic to p-harmonic. Assume M is oriented. Let $*$ denote the Hodge star operator. Define

$$*_p h := *|h|^{p-2}h.$$

Then $| *_p h|^{\frac{p}{p-1}} = |h|^p$, and

$$*_{\frac{p}{p-1}}(*_p h) = *^2 h = \pm h. \tag{2.12}$$

In particular, for h p-harmonic,

$$*_p h \in \mathcal{H}_{\frac{p}{p-1}}^{n-k}(M).$$

Interpreting Poincaré duality differently, we also note that the map $h \to |h|^{p-2}h$ defines a norm preserving bijection between $\mathcal{E}_p^k(M)$ and $\mathcal{H}_{p'}^k(M)$, where $p' := \frac{p}{p-1}$ denotes the dual exponent. Moreover, $z \to *z$ defines an isometric bijection between $T_p^k(M)$ and $\mathcal{T}_p^{n-k}$. This yields the following L_p-Poincaré duality theorem.

THEOREM 2.13. *Let M^n be a complete oriented Riemannian manifold. Then $*_p$ defines a bijective norm preserving map from*

$$\mathcal{H}_p^k(M) \quad to \quad \mathcal{H}^{n-k}_{\frac{p}{p-1}}(M).$$

Moreover, $h \to |h|^{p-2}h$ (respectively, $z \to *z$) defines a norm preserving bijection between $\mathcal{E}_p^k(M)$ and $\mathcal{H}_{p'}^k(M)$ (respectively, between $T_p^k(M)$ and $\mathcal{T}_p^{n-k}$).

See [**Pan08**, Corollary 14] for a proof of Poincaré duality for both reduced and torsion L_p-cohomology, not involving p-harmonic forms. For the convenience of the reader, we also give an elementary proof of the duality for the torsion part of the cohomology.

PROPOSITION 2.14. *$T_p^k(M) = 0$ if and only if $T_{p'}^{n-k+1}(M) = 0$.*

PROOF. By the closed image theorem [**BS18**, Theorem 6.2.3], the image of $d(C_p^{k-1}(M))$ is closed if and only if $d^*(\Omega_{p'}^k(M))$ is closed. Using $d^* = \pm * d*$, we see that $d^*(\Omega_{p'}^k(M))$ is closed if and only if $d(\Omega_{p'}^{n-k}(M))$ is closed. □

3. The Bochner formula

By elliptic regularity, a p-harmonic form h is smooth in a neighborhood of any point where it is nonzero. It is not necessarily smooth, however, at any point x_0 where $h(x_0) = 0$, but it is Hölder continuous. (See [**Uhl77**, Main Theorem].) We will not repeat that difficult proof here, but we note that a first step in proving the simpler result that $|h|$ is pointwise bounded involves extending the Bochner formula to p-harmonic forms. The usual vanishing theorems obtained via Bochner formulas for 2−harmonics then apply to the general p- harmonic case with no substantive change.

Given a differential form w, let $e(w)$ denote left exterior multiplication by w and let $e^*(w)$ denote the adjoint operation. Thus, for example, if $w = dF$ for some function F, then $e^*(w) = i_{\nabla F}$, where i_X denotes the interior product with X.

PROPOSITION 3.1 (Uhlenbeck [**Uhl77**, Theorem 1.10]). *Let h be a p-harmonic form. Then h satisfies the Bochner formula*

$$\Delta\frac{1}{2}|h|^p = -|h|^{p-2}\left(|\nabla h|^2 - (2-p)|d|h||^2 - \langle h, e(\omega^i)e^*(\omega^j)R_{ij}h\rangle\right)$$

$$- \frac{2-p}{2p}div\left([\langle e^*(\omega^k)\frac{h}{|h|}, e^*(\omega^j)\frac{h}{|h|}\rangle - \langle e(\omega^j)\frac{h}{|h|}, e(\omega^k)\frac{h}{|h|}\rangle](e_k|h|^p)e_j\right). \tag{3.2}$$

PROOF. Compute in an orthonormal frame:

$$- \Delta\langle h, |h|^{p-2}h\rangle = 2\langle\nabla_j h, \nabla_j|h|^{p-2}h\rangle - \langle\nabla^*\nabla h, |h|^{p-2}h\rangle - \langle h, \nabla^*\nabla|h|^{p-2}h\rangle$$

$$=2|h|^{p-2}|\nabla h|^2 - 2(2-p)|h|^{p-2}|d|h||^2 - \langle(dd^* + e(\omega^i)e^*(\omega^j)R_{ij})h, |h|^{p-2}h\rangle$$

$$- \langle h, (d^*d + e(\omega^i)e^*(\omega^j)R_{ij})|h|^{p-2}h\rangle$$

$$=2|h|^{p-2}|\nabla h|^2 - 2(2-p)|h|^{p-2}|d|h||^2 + \frac{2-p}{p}div(\langle e^*(\omega^k)\frac{h}{|h|}, e^*(\omega^j)\frac{h}{|h|}\rangle e_k(|h|^p)e_j)$$

(3.3)

$$- \frac{2-p}{p}div(\langle e(\omega^j)\frac{h}{|h|}, e(\omega^k)\frac{h}{|h|}\rangle e_k(|h|^p)e_j) - 2\langle h, e(\omega^i)e^*(\omega^j)R_{ij}|h|^{p-2}h\rangle.$$

Hence $|h|^p$ satisfies the claimed second order equation. □

Integrating the p-harmonic Bochner formula over M and applying the divergence theorem to eliminate the Laplacian term and the nonlinear divergence term yields the same corollaries as the usual 2− harmonic formula. Define the curvature operator

$$\mathcal{R} := e(\omega^j)e^*(\omega^k)R_{kj}.$$

COROLLARY 3.4. *If $\mathcal{R}$ is positive on k-forms, then $H^k_{p,red}(M) = 0$, for all $p \in (1,\infty)$. If $\mathcal{R}$ is nonnegative on k-forms, then any p-harmonic k-form must be covariant constant, and hence zero if the volume is infinite.*

COROLLARY 3.5. *If M is complete and Ricci flat with infinite volume,*

$$H^1_{p,red}(M) = 0,$$

for all $p \in (1,\infty)$.

COROLLARY 3.6. *If M^n has bounded sectional curvature and positive injectivity radius, there exists $C_M > 0$ such that $\|h\|_{L^\infty(M)} \leq C_M\|h\|_{L^p(M)}$, for every $h \in \mathcal{H}^k_p(M)$, $1 < p < \infty$, $0 \leq k \leq n$.*

PROOF. The Bochner formula (3.2) shows that $|h|^p$ is a subsolution to an elliptic equation of divergence form with uniformly bounded coefficients. The bounded sectional curvature and positive injectivity radius then allows one to use Moser iteration to control the sup norm of $|h|^p$ in terms of its average value on geodesic balls of radius less than or equal to inj_M. (See, for example, [**LS18**, Proposition 2.2] for the analogous $p = 2$ estimate.) This in turn implies a bound for $\|h\|_{L^\infty(M)}$ in terms of $(\mathrm{inj}_M)^{-\frac{n}{p}}\|h\|_{L_p}$. □

4. Monotonicity estimates for p-harmonic forms

The monotonicity estimates for p-harmonic forms can be derived exactly as the L^2 case treated in [**DS22**]. We repeat the computation here for the convenience of the reader.

Let $h \in \mathcal{H}^k_p(M)$. Let B_R be a geodesic ball in M with boundary S_R. Let r denote the radial function and ∂_r denote the radial vector field in B_R. Let $\{\lambda_a\}_a$ denote the eigenvalues of $\mathrm{Hess}(r)$. Choose an orthonormal eigenframe $\{e_a\}_a$ with dual coframe $\{\omega^a\}_a$, so that $e(\omega^i)e^*(\omega^j)\,\mathrm{Hess}(r)_{ji} = \sum_a \lambda_a e(\omega^a)e^*(\omega^a)$.

Define the functions $w_p(h,r)$ and $\mu_p(h,r)$ by

$$w_p(h,r) := \frac{\int_{S_r} |h|^p \sum_a \lambda_a(\frac{1}{p} - |e^*(\omega^a)\frac{h}{|h|}|^2)d\sigma}{\int_{S_r} |h|^p d\sigma}, \tag{4.1}$$

and

$$\mu_p(h,r) := \frac{\int_{S_r} |h|^p|e^*(dr)\frac{h}{|h|}|^2 d\sigma}{\int_{S_r} |h|^p d\sigma}. \tag{4.2}$$

PROPOSITION 4.3. *With w_p, μ_p, and h as defined above,*

$$\int_{B_\sigma} w_p(h,r)|h|^p dv = e^{-\int_\sigma^\tau \frac{pw_p(h,s)}{1-p\mu_p(h,s)}\frac{ds}{s}} \int_{B_\tau} w_p(h,r)|h|^p dv. \tag{4.4}$$

PROOF. We use the fact that on a Riemannian manifold, we can express the Lie derivative acting on forms in terms of the covariant derivative:

$$L_{\partial_r} = di_{\partial_r} + i_{\partial_r}d = \nabla_{\partial_r} + e(\omega^i)e^*(\omega^j)\,\mathrm{Hess}(r)_{ji}. \tag{4.5}$$

By Stokes's theorem we have

$$\int_{B_R} \langle L_{\partial_r} h, |h|^{p-2}h\rangle dv = \int_{S_R} |h|^{p-2}|e^*(dr)h|^2 d\sigma, \tag{4.6}$$

and by (4.5),

$$\begin{aligned}\int_{B_R} &\langle L_{\partial_r} h, |h|^{p-2}h\rangle dv \\ &= \int_{B_R} \langle(\nabla_{\partial_r} + e(\omega^i)e^*(\omega^j)\operatorname{Hess}(r)_{ji})h, |h|^{p-2}h\rangle dv \\ &= \int_{B_R} (\frac{1}{p}\frac{\partial |h|^p}{\partial r} + \langle e(\omega^i)e^*(\omega^j)\operatorname{Hess}(r)_{ji})h, |h|^{p-2}h\rangle) dv \\ &= \int_{B_R} (\frac{1}{p}|h|^p\Delta r + \langle e(\omega^i)e^*(\omega^j)\operatorname{Hess}(r)_{ji})h, |h|^{p-2}h\rangle) dv + \frac{1}{p}\int_{S_R} |h|^p d\sigma.\end{aligned} \tag{4.7}$$

Combining (4.6) and (4.7) yields

$$\int_{S_R} (\frac{1}{p} - \mu_p(h,R))|h|^p d\sigma = \int_{B_R} w_p(h,r)|h|^p dv. \tag{4.8}$$

Setting $Y(t) := \int_{B_t} w_p(h,r)|h|^p dv$, we rewrite (4.7) as

$$(\frac{1}{p} - \mu_p(h,R))Y' = w_p(h,R)Y(R). \tag{4.9}$$

Integrate this expression to obtain

$$\int_{B_\sigma} w_p(h,r)|h|^p dv = e^{-\int_\sigma^\tau \frac{pw_p(h,s)}{1-p\mu_p(h,s)}\frac{ds}{s}} \int_{B_\tau} w_p(h,r)|h|^p dv, \tag{4.10}$$

as claimed. □

PROPOSITION 4.11. *Let h be a p-harmonic k-form. Then with w_p and μ_p defined by (4.1) and (4.2), we have*

$$\lim_{r\to 0} rw_p(h,r) = \frac{n-1-pk+p\lim_{r\to 0}\mu_p(h,r)}{p}. \tag{4.12}$$

If $h(o) \neq 0$, then

$$\lim_{r\to 0} \mu_p(h,r) = \frac{k}{n}, \tag{4.13}$$

and

$$\lim_{r\to 0} rw_p(h,r) = (n-1)(\frac{1}{p} - \frac{k}{n}). \tag{4.14}$$

PROOF. For h smooth in a neighborhood of o, equation (4.13) is [**DS22**, Lemma 18]. By elliptic regularity, however, h is smooth in a neighborhood of any point at which it is nonzero, and (4.13) follows.

Hess(r) converges to the Euclidean Hessian of r, as $r \to 0$. Hence all λ_a approach $\frac{1}{r}$ and (4.12) follows. Finally, (4.14) is an immediate consequence of (4.13) and (4.12). □

COROLLARY 4.15. *If $p < \frac{n}{k}$ and $w_p(h,s) \geq 0$, for $s \in (0,R]$, then $\frac{1}{p} - \mu_p(h,s) > 0$, for $s \in (0,R]$. If $p > \frac{n}{k}$ and $w_p(h,s) \leq 0$, for $s \in (0,R]$, then $\frac{1}{p} - \mu_p(h,s) < 0$, for $s \in (0,R]$.*

PROOF. By (4.8) $\frac{1}{p} - \mu_p(h,s)$ does not change sign, given the hypotheses on w_p, and Proposition 4.11 fixes the sign. □

In order to exploit Proposition 4.3, we need geometric input. In the next section, we translate pinched negative curvature into lower bounds for w_p.

5. Negative sectional curvature

Let M^n be a nonnegatively curved n-manifold with $-1 \leq \sec_M \leq -\delta^2$, $\delta \geq 0$, where $\sec_M$ denotes the sectional curvature of M. Let r again denote the radial function of some geodesic ball, B_R, in M. By the Rauch comparison theorem [**Pet16**, Theorem 6.4.3], the eigenvalues $\{\lambda_a\}_a$ of $\mathrm{Hess}(r)$ satisfy

$$\delta\coth(\delta r) \leq \lambda_a \leq \coth(r), \tag{5.1}$$

(except for the zero eigenvalue associated to the radial direction) and $\mathrm{tr}\,\mathrm{Hess}(r)$ satisfies

$$\mathrm{tr}\,\mathrm{Hess}(r) \geq (n-1)\delta\coth(r). \tag{5.2}$$

In a neighborhood of some point $x_0 \in B_R \setminus \{0\}$, let $\{\omega^a\}_{a=1}^{n-1} \cup \{dr\}$ be a local orthonormal coframe such that the ω^a are metrically dual to Hessian eigendirections with eigenvalue λ_a. Let h be a k-form, with $|h|(x_0) \neq 0$. Let $J \subset \{1,\dots,n-1\}$ denote the indices of Hessian eigendirections such that at x_0, $|e^*(\omega^a)\frac{h}{|h|}|^2 > \frac{1}{p}$. Since h is a k-form, at x_0 we have

$$|e^*(dr)\frac{h}{|h|}|^2 + \sum_{a=1}^{n-1} |e^*(\omega^a)\frac{h}{|h|}|^2 = k. \tag{5.3}$$

Then

$$\begin{aligned}
\sum_a \lambda_a\Big(\frac{1}{p} - |e^*(\omega^a)\frac{h}{|h|}|^2\Big) &= \delta\coth(\delta r)\Big(\frac{n-1}{p} - k + |e^*(dr)\frac{h}{|h|}|^2\Big)\\
&\quad + \sum_a (\lambda_a - \delta\coth(\delta r))\Big(\frac{1}{p} - |e^*(\omega^a)\frac{h}{|h|}|^2\Big)\\
&\geq \delta\coth(\delta r)\Big(\frac{n-1}{p} - k + |e^*(dr)\frac{h}{|h|}|^2\Big)\\
&\quad + \sum_{a\in J} (\coth(r) - \delta\coth(\delta r))\Big(\frac{1}{p} - |e^*(\omega^a)\frac{h}{|h|}|^2\Big)\\
&\geq \delta\coth(\delta r)\Big(\frac{n-1}{p} - k\Big)\\
&\quad + (\coth(r) - \delta\coth(\delta r))\Big(\frac{|J|}{p} - \min\{|J|,k\}\Big)\\
&\geq \coth(r)\Big[\delta\Big(\frac{n-1}{p} - k\Big) + (1-\delta)\Big(\frac{k}{p} - k\Big)\Big]\\
&= \frac{k}{p}\coth(r)\Big[\delta\frac{n-k-1}{k} + 1 - p\Big]. \qquad (5.4)
\end{aligned}$$

This is positive for $p < \delta\frac{n-k-1}{k} + 1$ (which is less than $\frac{n}{k}$). In the quarter pinched case, $\delta = \frac{1}{2}$, and for $k \leq \frac{n-1}{2}$ $p < \delta\frac{n-k-1}{k} + 1$ specializes to $p < \frac{3}{2}$. Observe that the conjugate exponent to $\delta\frac{n-k-1}{k} + 1$ is $1 + \frac{k}{\delta(n-k-1)}$, (where by conjugate exponent, we mean the $p'(= \frac{p}{p-1})$ satisfying $\frac{1}{p} + \frac{1}{p'} = 1$).

For large p, we let L denote the eigendirections such that $|e^*(\omega^a)\frac{h}{|h|}|^2 < \frac{1}{p}$. Then

$$\begin{aligned}
\sum_a \lambda_a(|e^*(\omega^a)\frac{h}{|h|}|^2 - \frac{1}{p}) &= \delta\coth(\delta r)(k - \frac{n-1}{p} - |e^*(dr)\frac{h}{|h|}|^2) \\
&\quad + \sum_a(\lambda_a - \delta\coth(\delta r))(|e^*(\omega^a)\frac{h}{|h|}|^2 - \frac{1}{p}) \\
&\geq \delta\coth(r)(k - 1 - \frac{n-1}{p}) \\
&\quad + \sum_{a\in L}(\lambda_a - \delta\coth(r))(|e^*(\omega^a)\frac{h}{|h|}|^2 - \frac{1}{p}) \\
&\geq \coth(r)(n-k)\frac{p-1}{p}[\frac{\delta(k-1)}{n-k} - \frac{1}{p-1}]. \qquad (5.5)
\end{aligned}$$

This is positive for ($\frac{n}{k} <$) $\frac{n-k}{\delta(k-1)} + 1 < p$. For $\delta = \frac{1}{2}$ and $k \geq \frac{n+1}{2}$ this gives $3 < p$. Note that 3 is the conjugate exponent to $\frac{3}{2}$. More generally, the limiting exponents for k and $n-k$ are conjugate for fixed δ.

We have included the parenthetical comparisons to $\frac{n}{k}$ for easy application of Corollary 4.15.

Combining these positivity results with the estimates of the preceding section yields the following theorem.

THEOREM 5.6. *Let M be a complete Riemannian manifold with $-1 \leq \sec_M \leq -\delta^2 < 0$. Let h be a p-harmonic k-form on M. Let $\sigma \leq \tau \leq \text{inj}_M$. If $p < \delta\frac{n-k-1}{k} + 1$, then $w_p(h,r)$ is bounded below by a positive constant independent of h and r, and*

$$\int_{B_\sigma} w_p(h,r)|h|^p dv \leq e^{-(\tau-\sigma)k[\delta\frac{n-k-1}{k}+1-p]} \int_{B_\tau} w_p(h,r)|h|^p dv. \qquad (5.7)$$

If $p > \frac{n-k}{\delta(k-1)} + 1$, then

$$\int_{B_\sigma} w_p(h,r)|h|^p dv \leq e^{-(\tau-\sigma)k(p-1)[\frac{\delta(k-1)}{p} - \frac{(n-k)}{p(p-1)}]} \int_{B_\tau} w_p(h,r)|h|^p dv. \qquad (5.8)$$

These monotonicity estimates give us a new proof of Pansu's reduced L_p-cohomology vanishing theorem for simply connected complete manifolds of pinched negative curvature. The fact that we obtain precisely the same pinching constraints is not surprising as both proofs at some point rely on estimates for geometric quantities related to the eigenvalues of Hess(r).

COROLLARY 5.9 (Pansu). *Let M^n be a simply connected complete Riemannian n-manifold with $-1 \leq \sec_M \leq -\delta^2 < 0$. If $k < \frac{n}{p}$, and $p < \delta\frac{n-k-1}{k} + 1$, or $k > \frac{n}{p}$, and $p > \frac{n-k}{\delta(k-1)} + 1$, then*

$$\mathcal{H}_p^k(M) = 0 = H_{p,red}^k(M). \qquad (5.10)$$

PROOF. Let $h \in \mathcal{H}_p^k(M)$. Take the limit as $\tau \to \infty$ in Theorem 5.6 to deduce $\int_{B_\sigma} w_p(h,r)|h|^p dv = 0$, for all σ and all balls. Since $w_p(h,r) \neq 0$ for nonzero h and r sufficiently small, we obtain a contradiction unless $h = 0$. □

REMARK 5.11. In the $p = 2$ case, exponential bounds of the form given in Theorem 5.6 imply for M compact that $\frac{b^k(M)}{vol(M)}$ is $O(e^{-\mathrm{inj}_M k[\delta \frac{n-k-1}{k}+1-p]})$, and therefore imply the vanishing of the kth L^2 Betti number of M. The desire to exploit these bounds for such Betti number estimates was our initial motivation for extending these monotonicity estimates to $1 < p < \infty$. Unfortunately, more work remains to translate these estimates into Betti number bounds.

In the next section we extend the monotonicity estimates developed for p-harmonic forms to obtain vanishing theorems for the non-Hausdorff summand of the cohomology, even though that summand does not admit p-harmonic representatives. Once again we obtain the same pinching constraints as Pansu.

6. Closure of exact forms

In this section we consider the torsion $T_p^k(M)$. We give a new proof of a vanishing theorem of Pansu (see [**Pan08**]).

THEOREM 6.1. *(Pansu) Let M be a simply connected complete Riemannian manifold with $-1 \leq \sec_M \leq -\delta^2 < 0$. If $k-1 < \frac{n}{p}$, and $p < \delta\frac{n-k}{k-1} + 1$, then*

$$T_p^k(M) = 0. \tag{6.2}$$

PROOF. Let $\{z_j\}_{j=1}^\infty \subset B_p^k(M)$ be an L_p Cauchy sequence, with $z_j \overset{L_p}{\to} z \in Z_p^k(M)$. Let b_j denote the p-coclosed primitive of z_j. (See Lemma 2.2.) Repeating the first steps of the derivation of our monotonicity estimates given in Section 4, we have, since $db_j = z_j$,

$$\int_{B_R} \langle \{d, e^*(dr)\} b_j, |b_j|^{p-2} b_j\rangle dv$$
$$= \int_{S_R} |e^*(dr) b_j|^2 |b_j|^{p-2} d\sigma + \int_{B_R} \langle e^*(dr) z_j, |b_j|^{p-2} b_j\rangle dv,$$

and

$$\int_{B_R} \langle \{d, e^*(dr)\} b_j, |b_j|^{p-2} b_j\rangle dv = -\int_{B_R} w_p(b_j,r)|b_j|^p dv + \int_{S_R} \frac{1}{p}|b_j|^p d\sigma.$$

Combined, these yield

$$\int_{B_R} w_p(b_j,r)|b_j|^p dv \tag{6.3}$$
$$= \int_{S_R} (\frac{1}{p} - \mu_p(b_j,R))|b_j|^p d\sigma - \int_{B_R} \langle e^*(dr) z_j, |b_j|^{p-2} b_j\rangle dv.$$

Our pinching hypotheses imply $w_p(b_j,r) \geq \kappa_{k-1} > 0$, for some positive constant κ_{k-1}. Hence, taking the limit as $R \to \infty$, the S_R integral vanishes, and Hölder's inequality yields

$$\kappa_{k-1}\|b_j\|_{L_p} \leq \|z_j\|_{L_p}. \tag{6.4}$$

Since $\{z_j\}_{j=1}^{\infty}$ is Cauchy, $\{b_j\}_{j=1}^{\infty}$ is bounded. Passing to a subsequence, we can assume b_j coverges weakly in L_p to some b. Since d commutes with weak limits, $db = z$. Hence $\bar{B}_p^k / B_p^k = \{0\}$. □

7. Injectivity and pinched negative sectional curvature

Developing the techniques of the previous section, we next prove injectivity results for natural maps from $H^k_{p,red}(M) \to H^k_q(M)$, for $p < \delta\frac{n-k}{k-1} + 1$, $p < q$, and $q - p$ small.

By Corollary 3.6, p-harmonic forms with finite L_p norm are pointwise bounded on complete manifolds N with bounded sectional curvature and injectivity radius bounded below. Hence for $p \leq q$, $\mathcal{H}^k_p(N) \subset Z^k_q(N)$, and we can define a map

$$i^k_{p,q} : H^k_{p,red}(N) \overset{\text{Hodge isomorphism}}{\to} \mathcal{H}^k_p(N) \overset{\text{inclusion}}{\to} Z^k_q(N) \overset{\text{quotient}}{\to} H^k_q(N).$$

PROPOSITION 7.1. *Let M be a complete, simply connected, negatively curved n-manifold, with $-1 \leq \sec_M \leq -\delta^2 < 0$. If $p < 1 + \delta\frac{n-k}{k-1}$ and $\frac{q-p}{q} < \delta(n-k) - (p-1)(k-1)$, then $i^k_{p,q}$ is injective.*

PROOF. Let h be a p-harmonic representative of an element of the kernel of $i^k_{p,q}$. Fix a point $o \in M$, and let B_R denote the geodesic ball of radius R and center o. Let b_R denote the p-coclosed primitive of h restricted to $\bar{B}_R$. By (6.3) we have for $T \leq R$,

(7.2)
$$\int_{B_T} w_p(b_R, r)|b_R|^p dv = \int_{S_T} (\frac{1}{p} - \mu_p(b_R, T))|b_R|^p d\sigma - \int_{B_T} \langle e^*(dr)h, |b_R|^{p-2} b_R\rangle dv.$$

Suppose for $\epsilon := \frac{1}{2p}(\delta(n-k) - (p-1)(k-1) - \frac{q-p}{q})$, there exists a sequence $\{(T_j, R_j)\}_j$ with $T_j \to \infty$ and $R_j \geq T_j$, so that $\int_{B_{T_j}} |b_{R_j}|^p dv \leq \frac{1}{\epsilon^p} \int_{B_{T_j}} |h|^p dv$. Then passing to a subsequence, we can find $b_\infty \in L_p$ such that the b_{R_j} converge weakly to b_∞ on compact sets. Therefore $db_\infty = h$, which implies $h = 0$. Hence, for $h \neq 0$, $\exists T_0 > 0$ so that

$$\int_{B_T} |h|^p dv \leq \epsilon^p \int_{B_T} |b_{R_j}|^p dv, \forall T > T_0. \tag{7.3}$$

Setting $Y_R(T) := \int_{B_T} (w_p(b_R, r) - \epsilon)|b_R|^p dv$, we combine (7.2) and (7.3) to obtain for $T > T_0$,

$$Y_R(T) \leq \frac{(\frac{1}{p} - \mu_p(b_R, T))}{w_p(b_R, T) - \epsilon} \frac{dY_R(T)}{dT}.$$

Integrate this differential inequality from T_0 to R to get

$$Y_R(R) \geq e^{\int_{T_0}^{R} \frac{w_p(b_R, T) - \epsilon}{(\frac{1}{p} - \mu_p(b_R, T))} dT} Y_R(T_0) \geq e^{(R - T_0)[\delta(n-k) - (p-1)(k-1) - p\epsilon]} Y_R(T_0). \tag{7.4}$$

On the other hand, since $h \in Ker(i^k_{p,q})$, $\exists \beta_q \in L^q$ such that $d\beta_q = h$. Since b_R is the $L_p(B_R)$ norm minimizing primitive for $h_{|B_R}$, we have

$$\begin{aligned} \|b_R\|^p_{L_p(B_R)} &\le \|\beta_q\|^p_{L_p(B_R)} \\ &\le \|\beta_q\|^{\frac{p}{q}}_{L^q(M)} vol(B_R)^{\frac{q-p}{q}} \\ &\le \|\beta_q\|^{\frac{p}{q}}_{L^q(M)} vol(B_R)^{\frac{q-p}{q}} \\ &\le c(n)\|\beta_q\|^{\frac{p}{q}}_{L^q(M)} e^{\frac{q-p}{q}R}. \end{aligned} \tag{7.5}$$

As $R \to \infty$, (7.5) and (7.4) are mutually contradictory, unless $h = 0$. Hence the kernel of $i^k_{p,q}$ is $\{0\}$, as claimed. □

8. Symmetric spaces

The curvature pinching hypotheses in the preceding sections are somewhat coarse, as they depend on the Rauch comparison theorem. In more homogeneous geometries, we can often compute Lie derivatives of useful vector fields explicitly and can therefore achieve sharper results. As illustration, we resolve part of a conjecture of Gromov.

In this section, G is a semisimple Lie group of noncompact type, K is a maximal compact subgroup, and $X := G/K$ is the associated symmetric space. Let A be a maximal real split torus of G. (So $\text{rank}_{\mathbb{R}}(G) = dim(A)$.) Let $\mathfrak{g}$, $\mathfrak{k}$, and $\mathfrak{a}$ denote the Lie algebras of G, K and A, respectively. Let $\mathfrak{g} = \mathfrak{p} \oplus \mathfrak{k}$ denote the Cartan decomposition of $\mathfrak{g}$ determined by K.

CONJECTURE 8.1 (Gromov). *$H^k_p(X) = 0$, for $1 < p < \infty$, and $k < \text{rank}_{\mathbb{R}}(G)$.*

A choice of Weyl chamber determines a set Φ^+_G of positive roots of G relative to A. When G is not real split, extend $\mathfrak{a}$ to a Cartan subalgebra $\mathfrak{h}$ of $\mathfrak{g}$, and let $\hat{\Phi}^+_G$ denote the set of roots of $\mathfrak{h}$ whose restriction to $\mathfrak{a}$ lie in Φ^+_G. Let N be the unipotent subgroup of G whose Lie algebra $\mathfrak{n}$ is the direct sum of the root spaces with root in $\hat{\Phi}^+_G$. Let $\{\tilde{N}_\beta\}_{\beta\in\hat{\Phi}^+_G}$ be an orthonormal basis of $\mathfrak{n}$ with respect to the metric determined by the Killing form and the Cartan involution. Let $\{\tilde{\omega}_\beta\}_{\beta\in\hat{\Phi}^+_G}$ denote the dual basis. We identify elements of the Lie algebras of A and N with left invariant vector fields on A, and N, respectively, in the usual fashion. With this notation, we have the Iwasawa decomposition $G = NAK$, and X is isometric to $N \times A$ equipped with the metric

$$g_X := da^2 + \sum_{\beta\in\hat{\Phi}^+_G} a^{-2\beta}(\tilde{\omega}^\beta)^2,$$

where da^2 is a Euclidean metric on A. Orthonormalize our TN subframe with respect to this metric, setting $N_\beta := a^\beta \tilde{N}_\beta$, and $\omega^\beta := a^{-\beta}\tilde{\omega}^\beta$.

We now adapt our monotonicity machinery to the symmetric setting. Let T be a unit vector field in the Lie algebra $\mathfrak{a}$ of A. Then

$$\nabla_{N_\beta} T = -\beta(T)N_\beta. \tag{8.2}$$

Hence we can write the Lie derivative with respect to T acting on forms as

$$L_T = di_T + i_T d = \nabla_T - \sum_{\beta\in\hat{\Phi}^+_G} \beta(T)e(\omega^\beta)e^*(\omega^\beta). \tag{8.3}$$

For h p-harmonic,

$$0=\int_X \langle L_T h, |h|^{p-2}h\rangle dv = \int_X \sum_{\beta\in\Phi_G^+} \beta(T)|h|^{p-2}(\frac{1}{p}|h|^2 - |e^*(\omega^\beta)h|^2)dv. \tag{8.4}$$

PROPOSITION 8.5. *If there exists a unit vector $T \in \mathfrak{a}$ and some $\delta > 0$, so that for all k-forms h,*

$$\sum_{\beta\in\hat{\Phi}_G^+} \beta(T)(\frac{1}{p}|h|^2 - |e^*(\omega^\beta)h|^2) \geq \delta|h|^2, \tag{8.6}$$

then $\mathcal{H}_p^k(X) = 0$.

PROOF. Suppose T satisfying the hypotheses exists. Let h be a p-harmonic k-form. By (8.4), we have

$$\delta\int_X |h|^p dv \leq \int_X \sum_{\beta\in\hat{\Phi}_G^+} \beta(T)|h|^{p-2}(\frac{1}{p}|h|^2 - |e^*(\omega^\beta)h|^2)dv = 0. \tag{8.7}$$

Hence $h = 0$. □

The preceding proposition can be used to prove many vanishing theorems. We explore its application to Conjecture 8.1.

THEOREM 8.8. *Assume G is simple. Then $\mathcal{H}_p^k(X) = 0$, for $k < \mathrm{rank}_{\mathbb{R}}(G)$ and $1 < p \leq 2$.*

PROOF. The vanishing of $\mathcal{H}_2^k(X)$, for $k \neq \frac{\dim(X)}{2}$ is well known. See [**Bor85**], and more recently, see, for example, [**Olb02**]. So, we are left to treat the case $p < 2$. First assume G is real split. Let μ denote the maximal root in Φ_G^+. Let $T = T_G$ be the vector in $\mathfrak{a}$ which is metrically dual to $\frac{\mu}{|\mu|}$. Then for $\beta \in \Phi_G^+$, $\beta(T) \in \{0, \frac{|\mu|}{2}, |\mu|\}$, with $|\mu|$ only achieved for $\beta = \mu$. Let $n_G(1) := |\{\beta \in \Phi_G^+ : \beta(T) = 1\}|$, and $n_G(2) := |\{\beta \in \Phi_G^+ : \beta(T) = 2\}|$. Then $n_G(2) = 1$, for G real split. Hence

$$\sum_{\beta\in\Phi_G^+} \frac{\beta(T)}{p}|h|^2 = \frac{(n_G(1)+2)|\mu|}{2p}|h|^2. \tag{8.9}$$

On the other hand,

$$\sum_{\beta\in\Phi_G^+} \beta(T)|e^*(\omega^\beta)h|^2 \leq (k+1)\frac{|\mu|}{2}|h|^2. \tag{8.10}$$

Hence condition (8.6) is satisfied if $\frac{(n_G(1)+2)}{p} > (k+1)$. In other words, if

$$p < \frac{n_G(1)+2}{k+1}. \tag{8.11}$$

Case (1) For the A_n root system, in the usual notation, Φ_G^+ is given by $\lambda_i - \lambda_j$, $1 \leq i < j \leq n+1$, and $\mu = \lambda_1 - \lambda_{n+1}$. Hence $n_G(1) = 2n-2$ and condition (8.6) holds for

$$p < \frac{2n}{k+1}. \tag{8.12}$$

Case (2) For the B_n root system, $n \geq 2$, Φ_G^+ is given by $\lambda_i \pm \lambda_j$, $1 \leq i < j \leq n$, and λ_j, $1 \leq j \leq n$, and $\mu = \lambda_1 + \lambda_2$. Then $n_G(1) = 4n - 6$, and condition (8.6) holds for

$$p < \frac{4n-4}{k+1}. \tag{8.13}$$

Case (3) For the C_n root system, $n \geq 2$, Φ_G^+ is given by $\lambda_i \pm \lambda_j$, $1 \leq i < j \leq n$, and $2\lambda_j$, $1 \leq j \leq n$, and $\mu = 2\lambda_1$. Then $n_G(1) = 2n - 2$, and condition (8.6) holds for

$$p < \frac{2n}{k+1}. \tag{8.14}$$

Case (4) For the D_n root system, $n \geq 4$, the roots of N are given by $\lambda_i \pm \lambda_j$, $1 \leq i < j \leq n$, and $\mu = \lambda_1 + \lambda_2$. Then $n_G(1) = 4n - 8$, and condition (8.6) holds for

$$p < \frac{4n-6}{k+1}. \tag{8.15}$$

Case (5) For $G = G_2$, $n_G(1) = 4$, and condition (8.6) holds for

$$p < \frac{6}{k+1}. \tag{8.16}$$

Case (6) For $G = F_4$, $n_G(1) = 14$, and condition (8.6) holds for

$$p < \frac{16}{k+1}. \tag{8.17}$$

Case (7) For $G = E_6$, $n_G(1) = 20$, and condition (8.6) holds for

$$p < \frac{22}{k+1}. \tag{8.18}$$

Case (8) For $G = E_7$, $n_G(1) = 32$, and condition (8.6) holds for

$$p < \frac{34}{k+1}. \tag{8.19}$$

Case (9) For $G = E_8$, $n_G(1) = 56$, and condition (8.6) holds for

$$p < \frac{58}{k+1}. \tag{8.20}$$

For $k + 1 \leq \operatorname{rank}_{\mathbb{R}}(G)$, condition (8.6) thus holds in all cases for $p < 2$.

When G is simple but not real split, the restricted root system is one of the three classical systems of type A, B, or C, or the exceptional system F_4, but the multiplicity of the roots can increase, and for some short roots β, 2β may occur as a root. See [**Ara62**], especially pp. 32–33, for this and subsequent properties of restricted root systems. Then (8.9) and (8.10) become

$$\sum_{\beta \in \hat{\Phi}_G^+} \frac{\beta(T)}{p} |h|^2 = \frac{(n_G(1) + 2n_G(2))|\mu|}{2p} |h|^2, \tag{8.21}$$

and

$$\sum_{\beta \in \hat{\Phi}_G^+} \beta(T) |e^*(\omega^\beta) h|^2 \leq (k + \min\{k, n_G(2)\}) \frac{|\mu|}{2} |h|^2 \leq 2k \frac{|\mu|}{2} |h|^2. \tag{8.22}$$

When $n_G(2) > 1$, we often employ the simplified vanishing criterion

$$p < \frac{n_G(1) + 2n_G(2)}{2k} \tag{8.23}$$

instead of the sharper vanishing condition:

$$p < \frac{n_G(1) + 2n_G(2)}{k + \min\{k, n_G(2)\}}. \tag{8.24}$$

For a simpler comparison with our real split computations, let $n_{A_n}(j), n_{B_n}(j), n_{C_n}(j)$, $j = 1, 2$ denote $n_G(j)$ for G a real split simple Lie group of type A_n, B_n, or C_n, respectively. If G has restricted root system of type A_n, $n > 1$, and is not real split, then $n_G(2) \geq 2$, and $n_G(1) \geq 2n_{A_n}(1) = 4n - 4$. Hence condition (8.23) holds for

$$p < \frac{2n}{k}.$$

If G has restricted root system of type B_n, $n > 1$, and is not real split, then (8.23) holds for $p < 2$, since it holds in the real split case, and $n_{B_n}(1) + 2n_{B_n}(2) \leq n_G(1) + 2n_G(2)$. Similarly, (8.23) holds for $p < 2$, for all restricted root sytems of type F_4. So, we are left to consider the C_n case. Araki lists four different restricted root systems of type C_n, with

(1) $(n_G(1), n_G(2)) = (2n_{C_n}(1), 1) = (4n - 4, 1)$, and $\frac{n_G(1)+2n_G(2)}{k+1} = \frac{4n-2}{k+1}$.
(2) $(n_G(1), n_G(2)) = (4n_{C_n}(1), 3) = (8n - 8, 3)$, and $\frac{n_G(1)+2n_G(2)}{2k} = \frac{8n-2}{2k}$.
(3) $(n_G(1), n_G(2)) = (4n_{C_n}(1), 1) = (8n - 8, 1)$, and $\frac{n_G(1)+2n_G(2)}{k+1} = \frac{8n-6}{k+1}$.
(4) $(n_G(1), n_G(2)) = (8n_{C_n}(1), 1) = (16n-16, 1)$, and $\frac{n_G(1)+2n_G(2)}{k+1} = \frac{16n-14}{k+1}$.

In all these C_n instances, the vanishing criterion holds for $p < 2$, if $k < n$. □

The vanishing of the torsion part of the cohomology can be proved in a similar fashion.

THEOREM 8.25. *Let G be a simple Lie group. Then $T_p^k(X) = 0$, for $p < \frac{n_G(2)+2n_G(1)}{k-1+\min\{k-1,n_G(2)\}}$. In particular, $T_p^k(X) = 0$, for $k \leq \mathrm{rank}_{\mathbb{R}}(G)$ and $p \leq 2$*

PROOF. Let T_G be the vector field defined in the proof of Theorem 8.8. Let $\{z_j\}_{j=1}^{\infty} \subset B_p^k(X)$ be an L_p Cauchy sequence, with $z_j \overset{L_p}{\to} z \in Z_p^k(X)$. Let b_j denote the p-coclosed primitive of z_j. Then arguing as in Section 6, with T_G replacing the radial vector field, we have

$$\int_X \sum_{\beta\in\Phi_G^+} \beta(T_G)|b_j|^{p-2}(\frac{1}{p}|b_j|^2 - |e^*(\omega^\beta)b_j|^2)dv = -\int_X \langle i_{T_G}z_j, |b_j|^{p-2}b_j\rangle dv. \tag{8.26}$$

Hence $\|b_j\|_{L_p} \leq \frac{|T_G|}{\delta}\|z_j\|_{L_p}$, when

$$\sum_{\beta\in\Phi_G^+} \beta(T_G)|b_j|^{p-2}(\frac{1}{p}|b_j|^2 - |e^*(\omega^\beta)b_j|^2) \geq \delta|b_j|^2, \text{ for some } \delta > 0. \tag{8.27}$$

Arguing again as in Section 6, we deduce that $\{b_j\}_j$ is uniformly bounded in L^p and therefore admits a subsequence that converges to a primitive for z. Hence z is exact, and therefore $B_p^k(X)$ is closed. By (8.24), condition (8.27) holds when $p < \frac{n_G(2)+2n_G(1)}{k-1+\min\{k-1,n_G(2)\}}$. As we have recorded in the proof of Theorem 8.8, when $k \leq rank_{\mathbb{R}}G$, $p < \frac{n_G(2)+2n_G(1)}{k-1+\min\{k-1,n_G(2)\}}$ holds for $p \leq 2$, for all simple Lie groups, except when $p = 2$ for the real split Lie groups of type A_n and C_n. These can be handled by a Bochner type argument, which we include below. We treat the A_n case. The C_n case can be treated in the same manner.

Consider an $(n-1)$-form ϕ on $X = SL_{n+1}(\mathbb{R})/SO(n+1)$. Let t^j denote Euclidean coordinates on A, and let μ still denote the maximal root in $\Phi^+_{SL_{n+1}}$. Write

$$\begin{aligned}\sum_{\beta\in\Phi^+_G} \beta(T_G)(\frac{1}{2}|\phi|^2 - |e^*(\omega^\beta)\phi|^2) &= \frac{n_G(1)+2}{2}|\phi|^2 - |e^*(\omega^\mu)\phi|^2 - (n-1)|\phi|^2 \\ &\quad + \sum_{\alpha:\alpha(T_G)=0} |e^*(\omega^\alpha)\phi|^2 + \sum_j |e^*(dt^j)\phi|^2 \\ &= |\omega^\mu \wedge \phi|^2 + \sum_{\alpha:\alpha(T_G)=0} |e^*(\omega^\alpha)\phi|^2 \\ &\quad + \sum_k |e^*(dt^k)\phi|^2, \end{aligned} \tag{8.28}$$

which vanishes only if $\phi = \omega^\mu \wedge \phi_1$, where ϕ_1 is annihilated by all $e^*(dt^j)$ and all $e^*(\omega^\alpha)$, with $\alpha(T_G) = 0$. Combining (8.28) with the $p = 2$ case of (8.26) yields

$$\int_X \sum_k |e^*(dt^k)b_j|^2 dv = -2\int_X \langle i_{T_G} z_j, b_j\rangle dv. \tag{8.29}$$

Hence

$$\sum_k \|e^*(dt^k)b_j\|^2_{L^2} \le 2\|z_j\|_{L^2}\|b_j\|_{L^2}. \tag{8.30}$$

We will combine this with an estimate which is good when $\|e^*(dt^j)\phi\|^2_{L^2} < \epsilon^2\|\phi\|^2_{L^2}$, some small ϵ, in order to obtain estimates for the primitives, b_j. Expand

$$\nabla_{N_\beta} =: \nabla^0_{N_\beta} - \beta(\frac{\partial}{\partial t^j})(e(\omega^\beta)e^*(dt^j) - e(dt^j)e^*(\omega^\beta)), \tag{8.31}$$

where $\nabla^0_{N_\beta}$ maps the smooth span of $\{dt^j\}_j$ to itself and the smooth span of $\{\omega^\beta\}_\beta$ to itself. Write

$$d = e(dt^j)(\nabla_{\frac{\partial}{\partial t^j}} - \sum_{\beta\in\hat\Phi^+_G} \beta(\frac{\partial}{\partial t^j})e(\omega^\beta)e^*(\omega^\beta)) + e(\omega^\beta)\nabla^0_{N_\beta} =: d_A + d_N, \tag{8.32}$$

and

$$\begin{aligned} d^* &= -e^*(dt^j)(\nabla_{\frac{\partial}{\partial t^j}} - \sum_{\beta\in\hat\Phi^+_G} \beta(\frac{\partial}{\partial t^j})e^*(\omega^\beta)e(\omega^\beta)) - e(\omega^\beta)^*\nabla^0_{N_\beta} \\ &=: d^*_A + d^*_N. \end{aligned} \tag{8.33}$$

Then

$$\begin{aligned} \{d^*_A, d_N\} &= \{-e^*(dt^j)(\nabla_{\frac{\partial}{\partial t^j}} - \sum_{\beta\in\hat\Phi^+_G} \beta(\frac{\partial}{\partial t^j})e^*(\omega^\beta)e(\omega^\beta)), e(\omega^\alpha)\nabla^0_{N_\alpha}\} \\ &= -\sum_{\beta\in\hat\Phi^+_G} e^*(dt^j)\beta(\frac{\partial}{\partial t^j})e(\omega^\beta)\nabla^0_{N_\beta}. \end{aligned} \tag{8.34}$$

Expand

$$\begin{aligned}
&\|d\phi\|_{L^2}^2 + \|d^*\phi\|_{L^2}^2 \\
&= \|d_A\phi + d_N\phi\|_{L^2}^2 + \|d_A^*\phi + d_N^*\phi\|_{L^2}^2 \\
&= \|d_A\phi\|_{L^2}^2 + \|d_N\phi\|_{L^2}^2 + 2\langle\phi, \{d_A^*, d_N\}\phi\rangle_{L^2} + \|d_A^*\phi\|_{L^2}^2 + \|d_N^*\phi\|_{L^2}^2 \\
&= \|e(dt^j)(\nabla_{\frac{\partial}{\partial t^j}} - \sum_{\beta\in\hat{\Phi}_G^+} \beta(\frac{\partial}{\partial t^j})e(\omega^\beta)e^*(\omega^\beta))\phi\|_{L^2}^2 \\
&\quad + \|e^*(dt^j)(\nabla_{\frac{\partial}{\partial t^j}} - \sum_{\beta\in\hat{\Phi}_G^+} \beta(\frac{\partial}{\partial t^j})e^*(\omega^\beta)e(\omega^\beta))\phi\|_{L^2}^2 \\
&\quad + \|d_N\phi\|_{L^2}^2 - 2\sum_{\beta\in\hat{\Phi}_G^+} \beta(\frac{\partial}{\partial t^j})\langle\nabla^0_{N_\beta}e^*(\omega^\beta)\phi, e^*(dt^j)\phi\rangle_{L^2} + \|d_N^*\phi\|_{L^2}^2 \\
&= \sum_j \|\nabla_{\frac{\partial}{\partial t^j}}\phi\|_{L^2}^2 + \|\sum_{\beta\in\hat{\Phi}_G^+} e(dt^j)\beta(\frac{\partial}{\partial t^j})e(\omega^\beta)e^*(\omega^\beta)\phi\|_{L^2}^2 \\
&\quad - 2\langle\nabla_{\frac{\partial}{\partial t^j}}\phi, \sum_{\beta\in\hat{\Phi}_G^+} \beta(\frac{\partial}{\partial t^j})e(\omega^\beta)e^*(\omega^\beta)e^*(dt^j)\phi\rangle_{L^2} \\
&\quad + \|\sum_{\beta\in\hat{\Phi}_G^+} e^*(dt^j)\beta(\frac{\partial}{\partial t^j})e^*(\omega^\beta)e(\omega^\beta)\phi\|_{L^2}^2 \\
&\quad - 2\langle e^*(dt^j)\nabla_{\frac{\partial}{\partial t^j}}\phi, \sum_{\beta\in\hat{\Phi}_G^+} \beta(\frac{\partial}{\partial t^j})e^*(\omega^\beta)e(\omega^\beta))\phi\rangle_{L^2} \\
&\quad + \|d_N\phi\|_{L^2}^2 - 2\sum_{\beta\in\hat{\Phi}_G^+} \beta(\frac{\partial}{\partial t^j})\langle\nabla^0_{N_\beta}e^*(\omega^\beta)\phi, e^*(dt^j)\phi\rangle_{L^2} + \|d_N^*\phi\|_{L^2}^2 \\
&\geq \sum_j \|\nabla_{\frac{\partial}{\partial t^j}}\phi\|_{L^2}^2 + \|\sum_{\beta\in\hat{\Phi}_G^+} \beta(T_G)e(\omega^\beta)e^*(\omega^\beta)\phi\|_{L^2}^2 \\
&\quad + \|d_N\phi\|_{L^2}^2 + \|d_N^*\phi\|_{L^2}^2 - c_n(\|\nabla\phi\|_{L^2}\sqrt{\sum_j \|e^*(dt^j)\phi\|_{L^2}^2} + \sum_j \|e^*(dt^j)\phi\|_{L^2}^2) \\
&\geq \sum_j \|\nabla_{\frac{\partial}{\partial t^j}}\phi\|_{L^2}^2 + n^2\|\phi\|_{L^2}^2 + \|d_N\phi\|_{L^2}^2 + \|d_N^*\phi\|_{L^2}^2 \\
&\quad - c_n(\|\nabla\phi\|_{L^2}\sqrt{\sum_j \|e^*(dt^j)\phi\|_{L^2}^2} + \sum_j \|e^*(dt^j)\phi\|_{L^2}^2),
\end{aligned}$$

for some positive constant c_n. Specializing to $\phi = b_m$, with $d^*b_m = 0$ and $db_m = z_m$, we have

$$\begin{aligned}
&\sum_j \|\nabla_{\frac{\partial}{\partial t^j}}b_m\|_{L^2}^2 + n^2\|b_m\|_{L^2}^2 + \|d_N b_m\|_{L^2}^2 + \|d_N^* b_m\|_{L^2}^2 \\
&- c_n(\|\nabla b_m\|_{L^2}\sqrt{\sum_j \|e^*(dt^j)b_m\|_{L^2}^2} + \sum_j \|e^*(dt^j)b_m\|_{L^2}^2) \leq \|z_m\|_{L^2}^2. \qquad (8.35)
\end{aligned}$$

Combining (8.35) with (8.30) yields

$$\begin{aligned}&\sum_j \|\nabla_{\frac{\partial}{\partial t^j}} b_m\|_{L^2}^2 + n^2\|b_m\|_{L^2}^2 + \|d_N b_m\|_{L^2}^2 + \|d_N^* b_m\|_{L^2}^2\\ &\leq 2C_n(\|\nabla b_m\|_{L^2}\sqrt{\|z_m\|_{L^2}\|b_m\|_{L^2}} + \|z_m\|_{L^2}\|b_m\|_{L^2}) + \|z_m\|_{L^2}^2\\ &\leq 2C_n(\alpha\|\nabla b_m\|_{L^2}^2 + \frac{4+\alpha^2}{4}\|z_m\|_{L^2}\|b_m\|_{L^2}) + \|z_m\|_{L^2}^2. \qquad (8.36)\end{aligned}$$

Choosing α sufficiently small, we can absorb the $2C_n\alpha\|\nabla b_m\|_{L^2}^2$ term into the left-hand side to conclude that there is some constant $\hat{c}_n > 0$ such that

$$\|b_m\|_{L^2} \leq \hat{c}_n\|z_m\|_{L^2}.$$

The proof now proceeds as in the cases with $2 < \frac{n_G(2)+2n_G(1)}{k-1+\min\{k-1,n_G(2)\}}$. □

We remark that using the techniques of Section 7, we can use the vanishing of $\mathcal{H}_2^k(X)$, for $k \neq \frac{\dim(X)}{2}$ as a starting point for an alternate proof of (a slightly stronger version of) Theorem 8.8.

REMARK 8.37. In the papers [**BR20**], [**BR23a**], and [**BR23b**], Bourdon and Remy prove numerous vanishing and nonvanishing theorems for L_p-cohomology of symmetric spaces. Their vanishing results have significant intersection with those obtained above and also include vanishing results for ranges of k and p that we have not considered in this note. As in Pansu's work, their vanishing theorems are to a significant degree founded on the construction of primitives obtained by integration along orbits of flows. Our estimates utilize infinitesimal data associated with these same flows. Hence one expects significant overlap in the results reached by the two methods.

Acknowledgments

The author would like to thank the participants and the organizers of the AMS Special Session on the Singer–Hopf conjecture in geometry and topology in the March 2023 Spring Southeastern Sectional Meeting for stimulating and encouraging this work.

The author also thanks Luca Di Cerbo for numerous discussions on the contents of this note, and Marc Bourdon and Antonio Lopez for pointing out an error in an earlier draft of this paper.

References

[Ara62] S. Araki, *On root systems and an infinitesimal classification of irreducible symmetric spaces*, J. Math. Osaka City Univ. **13** (1962), 1–34. MR153782

[Bor85] A. Borel, *The L^2-cohomology of negatively curved Riemannian symmetric spaces*, Ann. Acad. Sci. Fenn. Ser. A I Math. **10** (1985), 95–105, DOI 10.5186/aasfm.1985.1012. MR802471

[BR20] M. Bourdon and B. Rémy, *Quasi-isometric invariance of continuous group L^p-cohomology, and first applications to vanishings* (English, with English and French summaries), Ann. H. Lebesgue **3** (2020), 1291–1326, DOI 10.5802/ahl.61. MR4191391

[BR23a] M. Bourdon and B. Rémy, *Non-vanishing for group L^p-cohomology of solvable and semisimple Lie groups* (English, with English and French summaries), J. Éc. polytech. Math. **10** (2023), 771–814, DOI 10.5802/jep.232. MR4585549

[BR23b] M. Bourdon and B. Remy, De Rham Lp-Cohomology For Higher Rank Spaces And Groups: Critical Exponents And Rigidity, Preprint, arXiv:2312.14025 [math.GR], (2024).

[BS18] T. Bühler and D. A. Salamon, *Functional analysis*, Graduate Studies in Mathematics, vol. 191, American Mathematical Society, Providence, RI, 2018, DOI 10.1090/gsm/191. MR3823238

[DS22] L. F. Di Cerbo and M. Stern, *Price inequalities and Betti number growth on manifolds without conjugate points*, Comm. Anal. Geom. **30** (2022), no. 2, 297–334, DOI 10.4310/cag.2022.v30.n2.a3. MR4516072

[Gro85] M. Gromov, *Asymptotic invariants of infinite groups*, Geometric group theory, Vol. 2 (Sussex, 1991), London Math. Soc. Lecture Note Ser., vol. 182, Cambridge Univ. Press, Cambridge, 1993, pp. 1–295. MR1253544

[ISS99] T. Iwaniec, C. Scott, and B. Stroffolini, *Nonlinear Hodge theory on manifolds with boundary*, Ann. Mat. Pura Appl. (4) **177** (1999), 37–115, DOI 10.1007/BF02505905. MR1747627

[KLV21] J. Kinnunen, J. Lehrbäck, and A. Vähäkangas, *Maximal function methods for Sobolev spaces*, Mathematical Surveys and Monographs, vol. 257, American Mathematical Society, Providence, RI, [2021] ©2021, DOI 10.1090/surv/257. MR4306765

[LS18] M. Lipnowski and M. Stern, *Geometry of the smallest 1-form Laplacian eigenvalue on hyperbolic manifolds*, Geom. Funct. Anal. **28** (2018), no. 6, 1717–1755, DOI 10.1007/s00039-018-0471-x. MR3881833

[Lu94] W. Lück, *Approximating L^2-invariants by their finite-dimensional analogues*, Geom. Funct. Anal. **4** (1994), no. 4, 455–481, DOI 10.1007/BF01896404. MR1280122

[Olb02] M. Olbrich, *L^2-invariants of locally symmetric spaces*, Doc. Math. **7** (2002), 219–237. MR1938121

[Pan08] P. Pansu, *Cohomologie L^p et pincement* (French, with English and French summaries), Comment. Math. Helv. **83** (2008), no. 2, 327–357, DOI 10.4171/CMH/126. MR2390047

[SS70] L. M. Sibner and R. J. Sibner, *A non-linear Hodge-de-Rham theorem*, Acta Math. **125** (1970), 57–73, DOI 10.1007/BF02392330. MR281231

[Sco95] C. Scott, *L^p theory of differential forms on manifolds*, Trans. Amer. Math. Soc. **347** (1995), no. 6, 2075–2096, DOI 10.2307/2154923. MR1297538

[Pet16] P. Petersen, *Riemannian geometry*, 3rd ed., Graduate Texts in Mathematics, vol. 171, Springer, Cham, 2016, DOI 10.1007/978-3-319-26654-1. MR3469435

[Uhl77] K. Uhlenbeck, *Regularity for a class of non-linear elliptic systems*, Acta Math. **138** (1977), no. 3-4, 219–240, DOI 10.1007/BF02392316. MR474389

Department of Mathematics, Duke University, Durham, North Carolina 27708
Email address: stern@math.duke.edu

Selected Published Titles in This Series

816 **Luca F. Di Cerbo and Laurenţiu G. Maxim, Editors,** Geometry and Topology of Aspherical Manifolds, 2025

814 **Susan Montgomery and Siu-Hung Ng, Editors,** Quantum Groups, Hopf Algebras, and Applications, 2025

812 **Ross Geoghegan, Craig R. Guilbault, and Kim Ruane, Editors,** Topology at Infinity of Discrete Groups, 2025

811 **Krishnendu Gongopadhyay and Shashikant A. Katre, Editors,** Geometry, Groups and Mathematical Philosophy, 2025

810 **Robert J. Berman and Yanir A. Rubinstein, Editors,** Convex and Complex: Perspectives on Positivity in Geometry, 2025

809 **Fabrice Baudoin and Luca Rizzi, Editors,** New Trends in Sub-Riemannian Geometry, 2025

808 **Loring W. Tu, Editor,** Group Actions and Equivariant Cohomology, 2024

807 **Galina Filipuk, Editor,** Recent Progress in Special Functions, 2024

806 **Maximilian Engel, Hildeberto Jardón-Kojakhmetov, and Cinzia Soresina, Editors,** Topics in Multiple Time Scale Dynamics, 2024

805 **Anthony Iarrobino, Pedro Macias Marques, Maria Evelina Rossi, and Jean Vallès, Editors,** Deformation of Artinian Algebras and Jordan Type, 2024

804 **Mahir Bilen Can and Jörg Feldvoss, Editors,** A Glimpse into Geometric Representation Theory, 2024

803 **Peter Gothen, Margarida Melo, and Montserrat Teixidor i Bigas, Editors,** Moduli Spaces and Vector Bundles—New Trends, 2024

802 **Ralph M. Kaufmann, Martin Markl, and Alexander A. Voronov, Editors,** Higher Structures in Topology, Geometry, and Physics, 2024

801 **K. A. Brown, T. J. Hodges, M. Vancliff, and J. J. Zhang, Editors,** Recent Advances in Noncommutative Algebra and Geometry, 2024

800 **Avinoam Mann, Louis H. Rowen, David J. Saltman, Aner Shalev, Lance W. Small, and Uzi Vishne, Editors,** Amitsur Centennial Symposium, 2024

799 **Alberto A. Condori, Elodie Pozzi, William T. Ross, and Alan A. Sola, Editors,** Recent Progress in Function Theory and Operator Theory, 2024

798 **Marat V. Markin, Igor V. Nikolaev, and Carsten Trunk, Editors,** Advances in Functional Analysis and Operator Theory, 2024

797 **Sangita Jha, Mrinal Kanti Roychowdhury, and Saurabh Verma, Editors,** Recent Developments in Fractal Geometry and Dynamical Systems, 2024

796 **John Cremona, John Jones, Jennifer Paulhus, Andrew V. Sutherland, and John Voight, Editors,** LuCaNT: LMFDB, Computation, and Number Theory, 2024

795 **Michael A. Jones, David McCune, and Jennifer M. Wilson, Editors,** Mathematical Analyses of Decisions, Voting and Games, 2024

794 **Jean-Pierre Magnot, Editor,** Recent Advances in Diffeologies and Their Applications, 2024

793 **Abba Gumel, Editor,** Mathematical and Computational Modeling of Phenomena Arising in Population Biology and Nonlinear Oscillations, 2024

792 **Shaoming Guo, Zane Kun Li, and Brian Street, Editors,** Recent Developments in Harmonic Analysis and its Applications, 2024

791 **Mee Seong Im, Bach Nguyen, and Arik Wilbert, Editors,** Algebraic and Topological Aspects of Representation Theory, 2024

790 **Peter Crooks and Alexandru I. Suciu, Editors,** Compactifications, Configurations, and Cohomology, 2023

For a complete list of titles in this series, visit the
AMS Bookstore at **www.ams.org/bookstore/conmseries/**.